PERSPECTIVES IN GEOMORPHOLOGY

VOLUME THREE

By the Same Author

- Climate Change: Bio-Diversity and Green Economy with *S. Padmaja* and *Ganesh Sharma*
- Indian Geomorphology
- Mathematical Modelling in Geographical Information System Global Positioning System and Digital Cartography with *D.R. Ram, Rama Prasad* and *P.R. Binda*
- Modeling in Resource Management and Environment Through Geomatics with *P.R. Binda*
- Ranthambhore Sanctuary: Dilemma of Eco-Development
- Ravine Erosion in India
- Sustainable Development: Issues and Case Studies with *Sri K. Chattopadhyay*
- Tropical Geomorphology: A Morphogenetic Study of Rajasthan

About the Editor

Prof. H. S. Sharma (*b.* 1945) holds post-graduate and doctoral degrees in Geography from the University of Saugar, Madhya Pradesh. Prof. Sharma is a former Professor and Head, Department of Geography, University of Rajasthan, Jaipur. He is also a former Director of the Indira Gandhi Centre for Human Ecology, Environment and Population Studies (HEEPS), Academic Staff College and Dean, Faculty of Science, University of Rajasthan.

Prof. Sharma has over 100 research papers in National and International journals like *Nature* and 16 books to his credit. He has supervised more than three dozen Ph.D.s. Prof. Sharma has travelled widely across U.K., U.S.A., Germany, Holland, Canada, Norway, Seychelles and Malaysia to attend and chair technical sessions in International Conferences and Exchange Programmes.

Prof. Sharma has successfully conducted several research projects sponsored by the U.G.C., Ministry of Environment and Forests, DRDO, New Delhi, NORAD, Norway and ISRO (SAC) Ahmedabad.

Prof. Sharma has been President of IIG, IGI and NAGI. He is also a President of the Commonwealth Human Ecology Council (India Chapter). Recently, he has been awarded Senior Fellowship by Indian Council of Social Science Research, New Delhi to work on Climate Change and its Impact on Land and Water Resources of Rajasthan.

Concept's International Series in Geography No. 2

PERSPECTIVES IN GEOMORPHOLOGY

VOLUME THREE

APPLIED GEOMORPHOLOGY

Edited by
H.S. Sharma

CONCEPT PUBLISHING COMPANY PVT. LTD.
NEW DELHI-110059

ISBN-13: 978-81-7022-123-4 (Set)

Edition 2016

Published and Printed by

Concept Since 1974

Concept Publishing Company Pvt. Ltd.
A/15-16, Commercial Block, Mohan Garden
New Delhi-110059 (India)
T: 25351460, 25351794, *F*: 091-11-25357109
E: publishing@conceptpub.com, W: www.conceptpub.com
Editorial Office: H-13, Bali Nagar, New Delhi-110 015, India.

Cataloging in Publication Data--Courtesy: D.K. Agencies (P) Ltd. <docinfo@dkagencies.com>

Perspectives in geomorphology / edited by H.S. Sharma.
volumes cm. -- (Concept's international series in geography ; no. 2)
Reprint. Originally published: 1981-1982.
Includes bibliographical references and index.
Contents: volume 1. Recent trends -- volume 2. Quantitative fluvial geomorphology -- volume 3. Applied geomorphology volume 4. Essays on Indian geomorphology.
ISBN 9788170221234 (set)

1. Geomorphology. 2. Geomorphology--India. I. Sharma, H. S. (Hari Shanker), 1945- editor. II. Series: Concept's international series in geography ; no. 2.

GB401.5.P47 2016 DDC 551.4 23

PUBLISHER'S NOTE

THIS IS the second set of volumes in our International Geography Series. It follows close upon the heels of the five-volume set on *Perspectives in Agricultural Geography* edited by Dr. Noor Mohammad. We take this opportunity at the very outset to acknowledge the co-operation we have received from the Editor Dr. Noor Mohammad and the numerous contributors from all over the world who have made Series One possible. The wide acclaim and the heartening response from our readers—researchers and students in the field of agricultural geography—have encouraged us to plan the publication of this second set entitled *Perspectives in Geomorphology* edited by Dr. Hari Shanker Sharma. Here again we would like to thank the Editor and the contributors from various countries of the world who have given a truly international perspective to these volumes. We are confident that this series will also be equally well-received all over the world.

The Concept Series in Geography is now an on-going one and we are very happy to announce that plans are afoot to make available a third set of volumes in the very contemporary subject of Urban Geography. This series is expected to run into fifty volumes and the publication programme, hopefully, will commence early in 1982. The response to our brochure has been very encouraging and more than 150 articles have already been received including several from overseas contributors.

With the publication of the initial volumes of this new series we hope to announce the subject of the fourth series so that both students and prospective contributors will receive information in advance.

In the next few years we hope in this manner, to provide geographers with a complete compendium of the latest advances in major areas in geography. Here, we might emphasize again, that the benefits that the students in this country as well as in other developing countries will obtain will be immeasurable.

Finally, we crave once again, the indulgence of our readers for any shortcomings in these volumes. We shall learn with experience.

PREFACE

IN THE last three decades there has been considerable progress in all branches of science. The science of geomorphology has also achieved a great deal during this time, and geographers, geologists, hydrologists, pedologists and specialists from other fields who have contributed to the study of geomorphology, have become aware of the comparative suddenness of the change within the context of land form studies. As a matter of fact, it is becoming virtually impossible for a single scholar to command all aspects of geomorphology. However, there has been increasing specialization at all levels on the one hand and a need for synthesising the results of these researches, to benefit the scholars working in the field of geo-science.

The present project which has been sponsored by the Concept Publishing Company, New Delhi is an humble attempt to disseminate the knowledge of concepts, theories, models and recent trends in geomorphology with reference to the present conceptual revolution in this field. With this ambition in view original research papers dealing with modern quantitative techniques were invited from geo-scientists from all over the world. The basic themes along with the tentative titles of the papers were sent to the contributors in various branches of geo-science. The contributors have responded well with original research papers.

Later on, these papers were thoroughly edited and the papers found worthy for publication were classified and arranged into four volumes.

The first volume of the series contains the papers dealing with the modern trends in geomorphology. This volume has been divided into five sections. Section one deals with the stream lines of geomorphic research, climatic and climato-genetic geomorphology, planation surfaces, granitic inselbergs and duri-crusts. Section two discusses Karst processes and land forms. Papers dealing with glacial and peri-glacial landforms have been included in Section three. Section four contains papers dealing with the coastal processes and land forms in tropical region. Section five contains the papers dealing with remote sensing and geomorphology.

As considerable amount of work has been done in fluvial geomorphology, therefore, volume two has been devoted to the papers dealing with the various aspects of quantitative fluvial geomorphology.

Volume three contains papers on applied geomorphology. Papers dealing with land classification, geomorphological mapping, terrain evaluation, and application of photo-interpretation and remote sensing in geomorphological mapping have been included in this volume.

Geo-scientists in India have also become aware and realised the significance of land form studies in their researches. As a matter of fact considerable amount of work has been done in various aspects of geomorphology. The fourth volume of the series contains papers on Indian geomorphology. It is earnestly hoped that these volumes will serve the purpose of enlightening the readers regarding the kind of research which is being taken up by modern geomorphologists today. As the editor of these volumes and, in addition a contributor, I count myself highly fortunate in my fellow contributors. All those whose papers are included in these volumes have prepared and supplied their material in the face of manifest difficulties—mainly the difficulties of time which pressed hard upon every busy academician. Most of my fellow contributors have patiently dealt with manuscripts and with queries relating to their papers. Those who sent their materials early have been tolerant of the subsequent delay, as late comers have tolerantly borne pressure from myself to complete their papers.

It would be ungenerous to conclude this introduction without acknowledging the guidance, encouragement and help of a number of senior scientists.

I am indebted to Professor Moonis Raza for his suggestions in making the outline of the project during my visit to the Centre for Regional Development, Jawaharlal Nehru University, New Delhi from January to April, 1980. I am highly thankful to Prof. Vaidyanadhan of Waltair University and Prof. Pramila

Kumar of Saugar University for their suggestions and encouragement. I am also thankful to Shri P.S. Dhinwa and Shri Ram Avatar Sharma for their assistance throughout the preparation of these volumes. Staff members of the Geography Department, University of Rajasthan, Jaipur, also deserve thanks for their co-operation and kindness in the preparation of this project. Special thanks are due to the publisher who has calmly suffered the repeated redistribution of authors throughout the world and who maintained effective contacts with myself. In addition I may suitably take this occasion to make a block statement of gratitude, on behalf of all the contributors, to the numerous friends and colleagues who have at various stages read part or all of the manuscripts and have supplied comments, criticisms and suggestions.

Department of Geography
Rajasthan University
Jaipur (India)

H.S. SHARMA

Independence Day
15 August 1981

CONTENTS

LIST OF CONTRIBUTORS

Naresh Bedi

Scientist, National Remote Sensing Agency, Secunderabad, India

Donald R. Coates

Professor, Department of Geological Sciences, and Environmental Studies, State University of New York at Binghamton, Binghamton, New York-13901

K. N. Das

Reader, Post Graduate Department of Geography, Bhagalpur University, Bhagalpur, India

K. K. L. Das

Lecturer, Post Graduate Department of Geography, Bhagalpur University, Bhagalpur, India

N. K. De

Reader, Department of Geography, University of Burdwan, West Bengal, India

Jaromir Demek

President, Ceskoslovenska Geograficka Spolecnost, PR. Ceskoslovenske Akademii Ved., Sekretariat : 66282 Brno, Mendlovo nam la

Johannes F. Gellert

Geograph Und Geologe, DDR-15, Potsdam, am, Helene-Lange-Strabe, 8, Fernruf-21347

D. K. Ghosh

Geologist, Geological Survey of India, Nagpur, India

G. P. Gupta

Post Doctoral Researcher, Department of Geography, Rajasthan University, India

Mieczyslaw Klimaszewski

Professor, Polish Academy of Sciences Institute of Geography and Spatial Organization Physical Geography Department, UL SW Jana 22 31-018 KRAKOW

Abhimanyu Mahto

Research Scholar, Department of Geography, Ranchi University, Ranchi, India

Sudershan Mukerjee

Research Officer, National Atlas and Thematic Mapping Organization, Calcutta, India

M. V. N. Murthy

Deputy Director General, Geological Survey of India, Western Region, Jaipur, India

Nageshwar Prasad

Post Graduate, Department of Geography, Bhagalpur University, Bhagalpur, India

R. K. Rai

Reader, Department of Geography, North Eastern Hill University, Shillong, India

B. L. Sharma

Lecturer in Geography, S. D. Government College, Beawar, (Rajasthan), India

Surendra Singh

Geomorphologist, Central Arid Zone Research Institute, Jodhpur (Rajasthan), India

LIST OF PLATES

6. Upstream face of excavated pit for block 1 and 2 showing weathering profile with kaolinised pockets (K) and partially decomposed crumbly patches (Cr) in completely weathered ground mass. Soil profile (D and S) is overlying the granite mass
7. Showing fresh granite rock with initiation of weathering (W) along cracks (C), left bank of Hasdo-Bango dam project, M.P.
8. Showing crumbly patches (Cr), decomposed ground mass (D), and soil (S) in a weathering profile of canal cut section of Banki dam project, Ambikapur
9. Showing junction of fresh rock (F), decomposed rock with crumbly patches (P), decomposed rock without any patch of original rock, Ghunghuta project, Ambikapur

Chapter Sixteen

1. Landsat MSS Image, Band 7, of a part of Andhra Pradesh, India, dt. 26-2-75
2. Landsat MSS Image, Band 7, of a part of Rajasthan India, dt. 14-2-75
3. Landsat MSS Image, Band 7, of a part of Rajasthan, India, dt. 3-10-79

LIST OF FIGURES

LIST OF TABLES

INTRODUCTION

THE HISTORY of geomorphological research bears the testimony of its important applications in the investigation, evaluation and classification of land as a resource. In this context geomorphology is closely concerned with the interrelations between landforms, type of rocks, nature of soil, vegetation cover and drainage, and its area of study embraces those of geology, geography, pedology, ecology, agriculture and civil engineering. In all these disciplines, because of the character and distribution of rocks, minerals, soil, vegetation or drainage, the application of geomorphological knowledge becomes an essential tool; as for instance in interpretation of aerial photographs, in soil and geological mapping or in natural resource surveys.

Within recent years geomorphology has made vast strides. The natural scientists from all over the world have come to realize that geomorphological researches are very helpful in studying varied aspects of natural environment. For example, in the USSR and Belgium, studies of landforms have been made to facilitate navigation; in Senegal, a geomorphological survey was conducted to investigate soil salinization process; in Canada, to explore the drifting of the St. Lawrence ice; in Australia and New Guinea, geomorphology is used in soil landscape and in integrated resource survey of undeveloped areas; and in the USA, landform studies are being utilized both in mineral exploration and in solving hydrological problems of western arid parts. All these studies demonstrate the number and diversity of applied geomorphology. Judging from the results obtained in these countries and for manifold promising aspects of future work, particularly with regard to resource surveys in arid and semi-arid regions of the world, it appears highly desirable to develop this newly emerged natural science.

As the world population continues to rise and as other natural resources of all types come to be increasingly exploited and some vital ones become scarcer, so man is becoming more and more aware of his natural environment and its finite limitations. In the early period of human development people were very dependent on their natural environment; each small isolated group sought the optimum environment and adopted its way of life to fit that environment. Mostly living within the natural ecosystem without doing its serious harm. This relationship has been called determinism; the environment determined the man's adaptation to it, and he himself exerted relatively little stress on the environment.

Today we live in a world that we have the power to modify it to an increasing extent, partly through a high level of technology and partly through a large increase in population number. Despite this technological power, however, the environment cannot be exploited and disturbed beyond a certain point without invoking the law of diminishing returns and other adverse results. Moonis Raza (1980) therefore, has rightly pointed out while developing the philosophy on the nature of ecosystem that the essentiality of man at nature is based on "man as a natural being and nature as a human reality. . ." Since the naturalization of man and the humanisation of nature proceeds simultaneously through man's interaction with nature, human history may be viewed as to continuation of, and a qualitatively new stage in natural history—the two constituting a differentiated unity. . . Nature comes within the orbit of development primarily as a constituent element transforms it, and himself gets transformed in the process." It could therefore be said that present is a time of neo-determinism in which it is becoming increasingly vital to live harmoniously with nature within the entire ecosystem, as members of it rather than to impose human modifications and exploitation on the ecosystem as outsiders beyond and above it. It is essential to work with nature and not against nature. To achieve this, however, it is also essential to understand nature and the processes that operate within the natural system (King 1976).

It is the geomorphologist's goal to achieve this. Geomorphologists can, therefore, play a very important part in helping to make wise use of bounties of nature in a wide range of fields. One of the most recent developments in geomorphology has been the great interest shown in environmental and applied geomorphology which is concerned with the impact of man on the landscape, and conversely the impact of the landscape in all its aspects of man.

A field of applied geomorphology in which man and natural processes have interacted for a very long time is in the exploitation of land for farming. I. Douglas (1968) and Langbein and Schumm (1958) have done considerable work on this aspect of geomorphology. The studies from various parts of the world have shown that the use of land for farming and the consequent destruction of the natural forests have greatly increased the amount of sediment transport in streams. Cuchlaine A.M. King (1976) has stressed that the excess sediment is derived from the slopes under cultivation or other types of human land use.

Another field of applied geomorphology is the study of the impact on geomorphic processes of man-made works, such as large dams, river diversions and similar measures. The work of Russian geomorphologists. Makkaveyev (1972) on the impact of large scale engineering works on the biotic and geomorphic processes exemplifies this aspect of geomorphology. The paper by D.K. Ghosh in this volume also signifies it. Geomorphological mapping and land classification are other methods which seek to record as objectively as possible in a geomorphologically meaningful way. The character of the land, has been developed along one line to produce the land unit system and land classification. Geomorphological mapping is, therefore, the central theme of geomorphological research.

The present volume III contains 18 papers of eminent scholars in the field of geo-science. This volume opens with the paper of M.V.N. Murthy, who by quoting Tom Dale and Vernon Gill Carter in "Top Soil and Civilization" as "Man whether civilized or savage, is a child of nature. He must confine his actions to certain natural laws if he has to maintain his dominance over his environment. When he tries to circumvent the laws of nature, he usually destroys the natural environment that sustains him. And when his environment to make wise use of bounties of nature in a wide range of fields. One of the most recent developments in geomorphology has been the great interest shown in environmental and applied geomorphology which is concerned with the impact of man on the landscape, and conversely the impact of the landscape in all its aspects of man.

A field of applied geomorphology in which man and natural processes have interacted for a very long time is in the exploitation of land for farming. I. Douglas (1968) and Langbein and Schumm (1958) have done considerable work on this aspect of geomorphology. The studies from various parts of the world have shown that the use of land for farming and the consequent destruction of the natural forests have greatly increased the amount of sediment transport in streams. Cuchlaine A.M. King (1976) has stressed that the excess sediment is derived from the slopes under cultivation or other types of human land use.

Another field of applied geomorphology is the study of the impact on geomorphic processes of man-made works, such as large dams, river diversions and similar measures. The work of Russian geomorphologists. Makkaveyev (1972) on the impact of large scale engineering works on the biotic and geomorphic processes exemplifies this aspect of geomorphology. The paper by D.K. Ghosh in this volume also signifies it. Geomorphological mapping and land classification are other methods which seek to record as objectively as possible in a geomorphologically meaningful way. The character of the land, has been developed along one line to produce the land unit system and land classification. Geomorphological mapping is, therefore, the central theme of geomorphological research.

The present volume III contains 18 papers of eminent scholars in the field of geo-science. This volume opens with the paper of M.V.N. Murthy, who by quoting Tom Dale and Vernon Gill Carter in "Top Soil and Civilization" as "Man whether civilized or savage, is a child of nature. He must confine his actions to certain natural laws if he has to maintain his dominance over his environment. When he tries to circumvent the laws of nature, he usually destroys the natural environment that sustains him. And when his environment

As has been referred earlier that geomorphology plays a significant role in environmental management, the study of D.K. Ghosh on geomorphology of Ravi Basin in northwestern Himachal Himalayas and environmental management is a systematic geomorphic study for terrain classification and their significance with the aid of aerial photos and surface characters in the districts of Chhamba and Kangra within the Ravi Basin. The paper presents a systematic account of terrain classification based on morpho-tectonic features and paleo-graphic history of the region. It has been stressed that the region calls for an environmental management relating to drainage basin, soil erosion and mass wasting and the planning and execution of storage and diversion structures.

Quarternary geomorphology has been studied very little specially in India. K.K.L. Das and K.N. Das in their study of alluvial morphology of the North Bihar Plain have attempted to describe the alluvial morphology and drainage characteristics of streams of North Bihar Plain. The characteristics of various topographic prominences like levees, chaurs, tals (dead and deferred course of evershifting courses of rivers) meander bends, marshes, bluffs, flood plains, and alluvial cones have been analysed. In the later half of the paper applied aspect of alluvial features and morphological characteristics which greatly influence the socio-economic and agrarian aspects of North Bihar Plain have been studied.

Settlements are topographic expressions of their arrange-ments. It is true with the rural settlements of the tribal areas. The study of R.K. Rai is an attempt in this direction who brings out the relationship between geomorphology and settle-ments in Meghalaya. The topography of Meghalaya is multicyclic. Geomorphologically it is more or less a plateau characterized by radial and rectilinear drainage patterns. Due to dissected topography and lack of land available for cultiva-tion, the settlements are of scattered and dispersed type. The scattered settlements are mostly found on hill tops while dispersed settlements are formed along hill slopes and river valleys. The study also attempts to delineate settlement zones.

Geomorphology not only plays an important role in settle-ment distribution but also in agricultural land use and taxonomy.

Geomorphology influences the type, growth and output of agricultural crops to a great extent. B.L. Sharma in his study on "Geomorphic control in agricultural Taxonomy—a case study of Rajasthan" has analysed physiographic characteristics including geology, drainage and soil type. He has correlated and analysed the cropping patterns in relation to terrain characteristics. On the basis of his study he has delineated crop combination regions of Rajasthan.

It has been stressed in a number of studies that any planning for a region should be conducted in harmony with a geomorphological features. Nageshwar Prasad in his study has analysed the role of geomorphology in regional planning taking a case study of Barakar Basin. He has stressed the problem of soil erosion and its measures have been suggested.

Glaciation has produced and distributed a variety of products and features of economic value to man. Knowledge of the processes that have produced such benefits is often essential for their successful use and exploitation. Glacial deposits range from exotic occurrences of diamonds and gold to more prosaic sand and gravel resources. Donald R. Coates in his paper attempts to describe the applications of glacial deposits in engineering, land use and other purposes. He has evaluated economic resources giving examples from USA and Europe.

As has been referred earlier that geomorphic study is very significant in dam foundations. D.K. Ghosh in his study has analysed the nature of granites and the processes of weathering in transforming the topography. The overall pattern of weathering in Central India is erratic and deep up to a depth of about 50 feet. Decomposition and further disintegration have formed clayey soils. Where the weathering process is complete, partial decomposition is characterized by the presence of crumble and corestone blocks in the pulverised ground mass.

Land classification forms of significant aspect of applied geomorphology. The identification and recognition of land units and their tabular description enables a complex region to be analysed into all its component areas of different characters or potentiality. The paper by N.K. De is a methodological attempt to classify land into different units. The study area covers about 78 square kilometres in northern Calabria, southern Italy. The methodology followed by the author in his study was similar to that of surveys carried out by Commonwealth Scientific and Industrial Research Organisation (CSIRO), Division of land research and regional survey, Australia with modifications. The author has analysed pre-field work methods and post-field methods and land complexes have been delineated.

All sections of the economy are linked up in a greater or lesser degree with the geographical environment including the configuration of the earth surface. A detailed geomorphological map providing an exact and measurable picture of the relief meets the requirements advanced by various fields of economy tending to a more rational civilization of landforms. The configuration of the earth's surface is of great interest to agriculture, settlement, communication, hydrologic, engineering, tourism, recreation, and management of resources. Mieczysolaw Klimaszewske's paper attempts to signify the importance of a geomorphological map for resource utilization. He has traced the development of geomorphological mapping and discussed the principles of the preparation of a detailed geomorphological map. The author has also analysed and described in detail the geomorphological map of the Polish Tatra, Mountains, which due to constraints arising out of multicolour printing has not been reproduced in this volume.

The unprecedented development of geomorphological mapping after World War II means a new quantitatively and qualitatively superior stage in the development of geomorphology as a science. In the last three decades geomorphologists gained considerable experience and passed successively over to the standardization of geomorphological maps on various scales within international frame. J. Demek's paper on geomorphological mapping—progress and problems deals with the present state of geomorphological mapping on world wide scale, mainly on the basis of activities of the I.G.U. Commission on Geomorphological Survey and Mapping in 1968-76. Several problems of geomorphological mapping have been discussed in this paper and has been stressed that no further progress in geomorphological mapping and accordingly in geomorphology as a science is possible at the present time without the standardization of the content and modes of representation of geomorphological maps on various scales on international levels.

Gellert in his paper has evolved the principles of geomorphological regionalisation and encoding of geomorphological characteristics for computer processing of data. He has emphasized that geomorphological regionalisation involves the delimitation, characterization and cartographical representation of regional units composed of similar structure of surface forms and forming regional units of various ranks and size orders.

His second paper "uniform system of the surface form morphogenesis and geomorphological classification and regionalization" is based on the widely recognised knowledge that the formation of earth surface forms, the morphogenesis, occurs due to the dialectic alternate efforts realised within the sphere of the earth's surface with the effects or overcoming the gravitation. This type of studies form the integral part of applied geomorphology.

Aerospace remote sensing techniques have opened new vistas for landscape analysis and planned development of regional economic resources. Naresh Bedi in his study has stressed the importance of this technique in applied geomorphic studies taking two examples of hydrogeomorphic analysis in Andhra Pradesh and desertification studies around Jodhpur in western Rajasthan. On the basis of these studies the author has tried to prove that applied geomorphic studies can be fruitfully carried out using landsat data.

Sudarshan Mukerjee in his paper has also discussed theory of remote sensing and application of this technique in geomorphological mapping.

It is beyond doubt that photo-geomorphological techniques play an important role in integrated natural resource. Surendra Singh's paper on Central Luni Basin in western Rajasthan is an attempt to analyse this aspect. The author in his study has divided the area into land resource units, resource inventory and resources of each unit have been enumerated.

The paper by G.P. Gupta deals with the drainage system and flood hazard in Bharatpur (Rajasthan). He has emphasised that in studying flood problem the rational analysis of drainage system is essential.

SUGGESTED READINGS

Douglas, I., 1968: Natural and man-made erosion in the humid tropics of Australia, Malaysia and Singapore, Landforms and Geomorphology—Concepts and history, *Benchmark Papers in Geology*, Vol. 28, Dowden, Hutchinson & Ross, Inc. Stroudsbury and Pennsylvania.

C.A.M. King., 1976 Landforms and geomorphology: concepts and history, *Benchmark Papers in Geology*, Vol. 28, Dowden, Hutchinson & Ross, Inc. Stroudsbury, Pennsylvania.

Langbein, W.B. and Schumm, S.A., 1958, Yield of sediment in relation to mean annual precipitation, landforms and geomorphology: concepts and history. *Benchmark Papers in Geology*, Vol. 28, 1976, Dowden, Hutchinson and Ross, Inc. Stroudsbury, Pennsylvania.

Makkaveyev, N., 1972: The impact of large engineering projects on geomorphic processes in stream valleys, *Soviet Geog. Rev. Transl.* 13 (6), 387-393.

Moonis Raza, 1980: The nature of ecosystem: A synoptic view, Indo-Soviet Symposium on rational utilisation of renewable natural resources held at New Delhi, Dec. 10-16.

ONE

M. V. N. Murthy

THE EARTH SCIENCES : THE NEED FOR STUDIES "AS IF PEOPLE MATTERED"*

Vasudhaiva Kutumbakam
(The Earth is one family)
Ancient saying of the Indian sages.

Is not your life itself a gracious gift
From God ? Were God to charge you interest
For the least of His generous gifts to you
Could you find the wherewithal to pay it ?
Is not the world a common treasury
Into which each man and thing in nature deposit
All they have for the maintenance of all ?
Does the lark lend you its song, and the spring
Its sparkling water ? Does the oak lend you
Its welcome shade, and the palm its honeyed date ?
Does the sheep give you the wool, and the cow her milk
For interest ? Do the clouds sell you their rain,
And the sun its warmth and light ? What would be
Your life without these and other gifts, which God,
Through Nature showers upon you for your good ?

—*Mirdad*

"Man, whether civilized or savage, is a child of nature. He must conform his actions to certain natural laws if he has to maintain his dominance over his environment. When he tries to circumvent the laws of nature, he usually

* Inaugural Address delivered to the Summer Institute on Recent Developments in Applied Geography, Department of Geography University of Rajasthan.

destroys the natural environment that sustains him. And when his environment deteriorates rapidly, his civilization declines".

—Tom Dale and Vernon Gill Carter in "Top Soil and Civilization".

THE TITLE of this paper "The Earth Sciences: The Need for Studies 'As If People Mattered' ", is in part borrowed from the well known book by E.F. Schumacker, "Small is Beautiful : A Study of Economics as if People Mattered". This highly provocative book has stimulated me, like others, into thinking on basic issues which are relevant not only to the world but are especially relevant to India. The key to his unique approach is in my opinion, the words "*as if People mattered*" I am afraid, all of us, especially scientists, rarely remember that people do matter. All our pursuits of knowledge are relevant only if the results may lead to development, "as if the people mattered", not only those who are our contemporaries, but the generations yet to come. I suspect that this is the consequence of specialisation and the absence of a general liberal education which should give us the right perspective—as citizens, as the privileged elite, and as specialists. While specialisation is inevitable with the rapid growth of the sciences which tend to subdivide into subdisciplines, our studies of "more and more of less and less" should not make us lose sight of the picture in its entirety, and also forget the very person, viz., Man, for whose benefit these studies are being pursued. Not only economics, but all other pursuits of knowledge have been guilty of this grave mistake of seeking knowledge and applying them "as if people did not matter at all".

The last quotation at the beginning of my address is also taken from Schumacker's book, and this graphically sums up the second point I want to make. I would like to stress that our studies in the Earth Sciences must be pursued not only as if people mattered, but as if Nature which sustains our very existence also mattered. Mirdad brings out this in his beautiful inimitable fashion in the second quotation which I have given.

Man, in whom Nature produced the highest evolved being has unfortunately become her very enemy ! In his shock-provoking book, "*Future Chock*", Alvin Toffler, quotes Kenneth Boulding, an eminent economist and imaginative social thinker: "The world of today, ... is as different from the world in which I was born as that world was from Julius Caesar's. I was born in the middle of human history, to date, roughly. Almost as much has happened since I was born as happened before". Toffler expands this statement : ". . . . . if the last 50,000 years of man's existence were divided into lifetimes of approximately sixitytwo years each there would have been 800 such lifetimes. Of these 800, fully 650 were spent in caves. Only during the last seventy lifetimes has it been possible to communicate effectively from one lifetime to another—as writing made it possible to do. Only during the last six lifetimes did masses of men ever see a printed word. Only during the last four has it been possible to measure time with any precision. Only during the last two has any one ever used an electric motor. And the overwhelming majority of all the material goods we use in daily life today have been developed, within the present 800th lifetime". The accelerative thrust in the rate of change, and the rate at which our natural resources are being consumed, has alarmed thinking men in other parts of the world, much more than in India. Schumacker warns (p. 26): ". . . the idea that unlimited economic growth, more and more until everybody is saturated with wealth, needs to be seriously questioned on at least two counts : the availability of basic resources and alternatively or additionally, the capacity of the environment to cope with the degree of interference implied". He stresses that we are eating up the capital and there are many who agree with his prophesy that unless we halt exploiting and polluting nature, the future of mankind is doomed. Unless Man realises his responsibility towards not only his fellow beings, of today and tomorrow, but to all of Nature—both animate and inanimate, he will bring massive destruction and disaster. Schumacker writes with concern (p. 56): ". . . .there is the immediate question whether 'modernisation' as currently practised without regard to religious and spiritual values, is actually producing agreeable results. As far as the masses are concerned, the result appear to be disastrous—a collapse of the rural economy arising tides of unemployment in town and country, and the growth of a city proletariat without nourishment for either body or soul".

Let me now explain how the theme I have chosen has a relevance to the Earth Sciences in which I include all studies concerning the Earth, including geology, geophysics, geography —-applied and basic, climatology, meteorology, oceanography, hydrology and a host of other disciplines which deal with the only planet of the solar system which has life in its diverse manifestation. If these disciplines are pursued discretely, without keeping Nature and Man in view, our priorities of studies will get distorted, and they will produce bizarre results which will do more harm than good. The tragedy of today's compartmentalised specialisation is that there is hardly any understanding of their mutual relevance. Thus the results are never evaluated taking a total view and there is no attempt to understand their significance in relation to the Earth. We also ignore the effect of these studies in our plans and priorities of programmes for what we call development of the humanbeing and society, who not only fund our studies, but also, quite naturally expects useful results. By our fragmentary approach and pursuits, forgetting the total picture we are creating problems and unless we have an approach in our studies "as if people mattered". I am afraid we have failed in our responsibility to society.

Let me take the question of floods as an example since we are already reading about them in Assam where the Brahmaputra is reaching danger levels, and very scon we will have the story repeated elsewhere, once the monsoon is in full force. Anybody who knows the history of a river system is well aware that it has its origin in the hills, it flows downhill cutting gorges and is joined by tributaries and suddenly debauches into areas with relatively flat, wide valleys, and finally merges in the sea where it builds deltas. Flooding of the lower reaches is a natural phenomenon. This happens when the river valley, which is wide and not very deep in that is unable to confine the sudden additions to the volume of water when there is heavy rains in the source areas in the hills. When the population was small, floods were a blessings since the flood waters brought rich alluvium which in due course became good agricultural land with the growth of humus. But with the increase in population at an exponential rate, the situation has completely changed, especially during the last fifty years. Not only has the area of flooding enlarged but the frequency of floods has increased.

Let us now see how the problem has been tackled. The engineer has built several kilometres of defensive walls or embankments on either side of the present channel of the river to contain the flood waters and also protect the crops and the people who have with confidence settled in the immediate neighbourhood. The building of embankment interferes with the natural process of spreading of the flood waters and the deposition of the silt and alluvium over a large stretch. Now the silt settles in the river bed and within the narrow confines of the restricted river-channel, the limits now being defined by the embankments. Over the years the river bed has thus risen through accretion, and the level of the bed of the river is now higher than the surrounding areas beyond the embankments. The situation has become reversed. Man-made embankments are not always impregnable and we are dealing with a dynamic situation which changes cumulatively over the years. Thus there are always breaches in the embankments and desperate attempts to plug the breaches. This danger of breaches is also great when only the present course of the river has been taken into account by the engineer who is not naturally aware of the history of the migration of the rivers and also the nature of material during vertical accretion during alluviation.

During breaching, the areas which are at a level lower than the river bed get flooded and this water cannot be drained and usually forms swamps not only ruining the land permanently but also becoming a major health hazard as the breeding ground of mosquitoes. A solution similar to that tried years ago in the USA has now resulted in this kind of situation for most of the rivers in the Ganga-Yamuna alluvial plains and the Brahmaputra Valley. This man-made disaster is the result of partial solutions taking a partial view of the problem. I speak from personal experience since I was in Assam for six years from 1966 and was connected with a project initiated by me when we carried out geomorphic studies of the river systems. We cannot understand the problems in their totality unless we know the history of the rivers how they have migrated and the factors controlling their migration. We need not only geomorphic maps of the river system as it exists today, but also maps of successive periods. Older topographic maps prepared by the Survey of India are useful, but aerial photographs taken at intervals are most important, and provide a wealth of data which maps cannot give. These studies give us basic data and understanding of the river morphology and the factors that have governed its evolution. In a tectonically active area like Assam, we cannot ignore earthquakes which have affected the region in 1897, 1930 and 1950. The maps or photographs with the dates when they were prepared/taken have been used to evaluate the history of migration of the rivers and also the silting up. Geomorphic mapping has to be supplemented by sedimentological studies of the different surfaces and model studies are also vital. Without such basic data, I am afraid we will not be able to have a scientific basis for any rational solution of either the problem of flooding or of the proper management of the river—land system including proper land use.

Embankments have also other disastrous effects. Confined water flowing at a level higher than the surrounding level also tends to disturb the groundwater system. Since we now seem to think that the panacea for all our food problem to get additional crops and drought, is by full "exploitation" of our groundwater resources, we should be aware that we can play havoc with the natural system of this resource. Already over-exploitation of groundwater in the coastal parts of Gujarat has resulted in the incursion of seawater which has ruined the groundwater whose salinity has increased. Only the other day I read in the "Patriot" an article which says, how many wells have run dry in Haryana and also have use of artificial fertilisers, the groundwater has been ruined since the excess unused fertilisers have now polluted the waters. Unfortunately, unless we know how much water we can really use, without revising the system we will be in such dangers since Nature cannot keep pace with our greedy "thirst" in sucking up water faster than the natural rate of replenishment. We are killing the hen that lays the golden egg.

Climate has its effect on the natural system. In monsoonic climate as in Assam, the effect is different but as we move westwards aridity increases even in West Bengal, reaching its peak in western India including Rajasthan. The embankments and the canal systems which have been built up in Punjab, Haryana and in Uttar Pradesh have added new dimensions to our problems, created by our irrational, unscientific interference with the natural system of land and water (both surface and underground). Water always tends to dissolve salts which in an arid climate comes up to the surface by capillary action and the soil becomes alkaline and millions of acres of land have been rendered useless by this alkalinisation of soil. This is again the result of a fragmentary approach and also tackling problems without the requisite scientific data on the drainage system. Water has to drain away the salts. This needs sub-surface data since such drainage will depend on natural systems of aquifers. This needs not only geomorphic maps of the surface but also of the sub-surface and also sedimentological characteristics of the layers. It will be wise for us to carry out all these studies, now that the Rajasthan canal project is coming up in a big way and hopes of major developments have been raised. I need only say that the 'handwriting on the wall' is written in bold letters through the experience of several decades in our immediate neighbourhood and we should read the handwriting on the wall. We need to look at the totality of the picture because, Nature is one integrated system, to use the word in common usage and any interference with any one segment or parameter will create disequilibrium and will upset other parameters and upset the system as a whole and create other problems. We might have solved one problem for some time, but would have created additional problems some of which will need enormous effort to solve. I must hasten to add that all this is not the sole irresponsibility of either the politician, or the administrator or the engineer. The point I am making is that it is the responsibility of the Earth Scientists first to learn to think not merely in terms of their narrow specialisations but also view the problems in their totality, and fix priorities of their research taking into consideration problems which are vital to "people as if they mattered" rather than pursue research which will bring acclaim in academic circles in India or abroad, and lastly, and perhaps most importantly, have the courage of conviction to tell the truth and nothing but the truth so that the persons who matter in policy-making are aware of the dangers inherent in indiscriminate interference with Nature in our desperate hurry to solve immediate problems without the proper scientific basis.

I am reminded in this connection of an International Conference I attended at Saugar on "Deccan Traps and Plateau basalts of the world". This was attended by several scientists from India and some from abroad. At the concluding session there was unanimity on the urgent need for studies of the Deccan traps, their petrology, mineralogy, petrochemistry, etc. We recommended that we need at least 100 scholars to undertake this study. You know what happens when any group of people belonging to any profession meet, convinced that the only way is to increase the output of their particular branch of studies. At the end of the session we had an excursion outside Saugar and during lunch break I was in a group wherein there was an English expert on basalts who had worked in East Africa. He started all of us by asking the question; "Don't you think that in the present context of India's needs it is more important to putmore scientists in a systematic study and survey of groundwater than study the academic problems of basalts and their petrochemistry ?". We were all dumbfounded, and with a mouthful of juicy lunch it was easy not to answer his question. His question recurs in my mind repeatedly; I have often felt ashamed that it had to be asked by a foreigner, perhaps only a foreigner could ask that question for he could see the absurdity and irrelevance of our over-enthusiasm for Deccan traps and their petrochemistry. He was in short saying: Don't you think that your plans and priorities must be "as if people mattered"?

I have included as an appendix to this address a very interesting letter addressed to Mr. Franklin Pearco, President, U.S.A. by Chief Seattle, a chief of the Red Indian tribe in reply to the proposal that the American Government would like to "buy" the land occupied by the tribe, in return for 'reservation' and 'protection'. It is a remarkable letter which reveals basic differences in the attitude to Nature and Life.

The Chief asks : "How can you buy or sell the sky, the warmth of the land ? The idea is strange to us. Yet we do not own the freshness of the air, or the sparkle of the water. How can you buy them from us ? Every shining pine-needle, every sandy shore, every mist in the dark woods, every clearing, every humming insect is holy memory and experience of my people. . . . . . We know that the white man does not understand our ways. One portion of land is the same to him as the next. For he is a stranger who comes in the night, and takes from the land whatever he needs. The Earth is not his brother : but, his enemy and when he has conquered it, he moves on. . . . . ." The chief then describes the white man's cities : "There is no quiet place in the white man's cities— no place to hear the leaves of spring, the rustle of insect wings. But perhaps I am a savage and do not understand; the clatter only seems to hurt the ears. And what is there in life if a man cannot hear the lovely cry of the whippoorwill or the argument of frogs around a pond at night ? . . . . . The air is precious to the red man, for all things share the same breath—the beasts, the trees and the man. The white man does not seem to notice the air he breathes, like a man dying for many days, he is numb to the smell". Let me add that my intention of quoting extensively from the Red Indian Chief's letter is not to stress racial problems; I feel the Red man's letter is addressed to humanity as a whole, to all of us, who are deep in this game of 'exploiting' Nature. I can also hear the plaintive cry of rural India looking at urban India where lives the modern Indian version of the white man about whom the Chief speaks. It is significant that this uneducated Red Indian had in 1855 foreseen the disasters about which modern man is crying hoarse after making a complete mess of environment !

The Red Indian Chief's letter is a good introduction to the problems of pollution about which I would like to speak in the next few minutes. 120 years later, Dr. Lawrence Peter in his excellent book *Peter Plan* writes (p. 20-21) : "The first seventy-five years of this century have been witness to perpetuation of growth and greed with almost universal indifference to any consequence other than mere profit. . . . . .Man increased the ways in which he could work his will on nature. The methods that drew the shortest lines to the highest profit were the greatest despoilers. . . . . . . The pollutants are concentrated over the cities where the people are trapped in the thickening air. In Japan, the school children are bussed from the industrial cities so that they can see what clear air looks and feels like. In Italy, during the worst of the Milan winter smog season, travellers are advised to take the train rather than the plane; frequently pollution causes such poor visibility that the planes cannot land." Dr. Peter (p. 22-23) comes to some very drastic conclusions, reminiscent of Schumacker's theme : "Our traditional belief in the value of growth in all areas is based on the concept that bigger is better, and before we can measure our overall progress as a nation by our tabulation of our total GNP (gross national product). It is concerned only with quantity. . . . . . The increasing GNP. . . . . could be more production, a declining quality of goods and services a grossly deteriorating environment and escalating the pathological side-effects of growth for growth's sake. . . . The GNP is not a measure of the quality of life but merely the cost of goods and services. . . . The GNP could stand for Gross National Pathology". The situation in India is no better. New Delhi greets us with three chimneys which dominate the skyline puffing poison into the air; the Yamuna, we are assured will be completely polluted in 30 years. The less said about Calcutta's problems of pollution the better. In Bombay nothing can live in the highly polluted waters. Other cities and towns are racing ahead to get these fruits of industrial civilization.

All of us, seientists, economists, and technologists are fond of using the term 'exploitation' of natural resources. The dictionary defines the word exploitation as 'selfish utilization; utilization for profits'. How very true, and the fact that we use it liberally and quite naturally shows that we are not even ashamed of what we are doing. We are so proud of man's scientific and technological achievements in the full "exploitation" of our resources, which actually means "utilization of our resources to meet selfish ends without any thoughts of the consequences".

By the word pollution we always tend to think in terms of air and water. The term also means debase and degrade, which means reduce in quality. While the problems of pollution of air and water is well recognised, there is another aspect of this "pollution" and "degradation" about which I would now like to say a few words. I mentioned in the earlier part of this address the problems of floods and siltation of the river bed. The problem of floods is directly related to the problem of soil erosion which is taking place on an unprecedented scale largely due to exponential growth of population and the consequent problems of finding land to live on to meet his food needs. In-roads into virgin land and forests are being made; vegetation which holds and binds the soil, for agriculture is cleared and trees are felled indiscriminately for fuel and for building houses to live in. I would like to introduce you to a remarkable book which should be read by anyone concerned with the problem of land-water and food, a book by Erik P. Eckholm (1978), entitled "Losing Ground"; it is the result of studies made by him in different parts of the world including Nepal. Maurice F. Strong, Executive Director, United Nations Environment Program sums up beautifully the essence of the book in his foreword : "This new publication surveys and analyses on a systematic basis the various ways in which our delicately balanced food systems are being ecologically undermined. It provides concrete examples of the serious ecological deterioration that is taking place through deforestation, overgrazing, soil erosion and abandonment, desertification, the silting of irrigation systems and reservoirs and the changes in the frequency and severity of flooding. In all these, the increasing intensity of humanity's pressure on the land and the continuation of careless short-sighted land use practices compound the effects of such natural phenomena as droughts and floods; often turning the temporary problem they create to large scale disaster". . . . "Losing Ground. . . . . goes to the heart of the issue of human survival and it makes it clear that the issue will be decided by what people and their Governments do—or fail to do—at this point in our history. We seem to forget that soil is also a natural resource, and must be treated as such. In fact it is the most vital resource on the preservation of which will depend man's future on this planet. The rate of soil formation takes thousands of years and is the result of the nature of the rock type, the climate, and the vegetation which provides humus and acids for chemical alteration of the minerals of rocks. It is a complex process which we cannot induce once the soil cover is gone and the bare rock—the delight no doubt of the geologists —is exposed.

When the ship is sinking in the sea it sends out distress signals, "SOS . . . SOS . . . SOS" which stands for "Save Our Soul". Edgar D. Mitchell the sixth man to land on the moon saw the Earth from space and says in his Introduction entitled "From Outer Space to Inner Space" to the book "Psychic Exploration : A Challenge for Science": "The first thing that came to mind as I looked at Earth was its incredible beauty. . . It was a majestic sight—a splendid blue and majestic jewel suspended against a velvet black sky . . . Buckminster Fuller's description of the planet as " 'Spaceship Earth' seemed to be eminently fitting". SOS . . . SOS . . . SOS is the distress signal which Spaceship Earth is sending out: "Save Our Soul if We have to Save our Soul".

Eckholm has lucidly discussed the various aspects of the problem of soil erosion and also stressed about the cost factors. He concluded (p. 185) : "What is now needed however, is another giant step beyond such assessments to the incorporation of an ecological perspective into the development—planning process from its inception. A planning exercise with the natural environment's capacity to serve human needs as its reference point would, in many countries generate a different mix of priorities and projects than those supported by present day systems. The need is not just for an agency to predict detrimental environmental consequences of projects chosen on the basis of traditionally quantified financial variables, but is, rather, for one to identify the programmes and strategies needed to enhance the environment's ability to support an improved life for the people. This is after all, the ostensible goal of development". On p. 187 he says : "Seldom is a direct connection drawn between rural degeneration and mushrooming of urban shanty towns, where degraded human conditions are creating a special tinderbox and stand out as one of the major social blights of the closing country. Even more seldom is that critical connection acted upon. Providing a decent, sustainable habitat for more humans in the countryside is a far more practical and socially desirable than shouldering the almost insuperable task of providing a healthy habitat for burgeoning numbers in the cities, where no chance exists for productive universal employment. The tide of refugees can be slowed down only through rural ecological regeneration; when toposoil washes down the mountain sides, those whose livelihoods depend on it lose their foothold".

Now the question naturally arises how we are going to tackle the complex problems in a system which has degraded and also whose regeneration and return to an equilibrium seems to be impossible. Fortunately we have various means at our disposal, for performing this 'miracle', provided we use these methods judiciously, provided we cooperate with one another subordinating our narrow outlook and interests and work towards the common end, "as if people mattered". We in the GSI have made a beginning and we have gained some experience in the preparation and utilization of various types of geological, geomorphic, geohydrological soil maps, in collaboration with various sister agencies we have almost completed such work in the Ananthapur district in Andhra Pradesh, are half way through in Dharmapuri district in Tamil Nadu. We have also studied the rural-urban complexes in Visakhapatnam, Andhra Pradesh, the Bangalore region and Hyderabad-Secunderabad. During the past two years we have covered Jaipur and its surrounding rural areas, and also Ajmer-Pushkar and the surrounding areas. As I already mentioned earlier the maps prepared in connection with the studies of the flood problem in Assam-West Bengal provide the basis for rational land-water use. The several maps must form overlays, and we also need synoptic maps to understand the mutual relation between the various natural parameters of the sub-systems forming the total system. In our natural anxiety for rural development—there can be no two opinions on this—we are forgetting the relation between rural and urban areas.

Fortunately with the availability of aerial photos and imagery we can prepare these maps much faster than by traditional methods. We need plans and strategies based on scientific data. Only maps can provide such data and information and not statistics which now forms the basis. Statistics by itself will not help in formulating plans and strategies. Statistics can also be cooked up, but maps cannot be. Aerial photographs can always be used for checking up.

However, I would like to add that all these studies will have to be carried out in an integrated fashion. Studies may have to be carried out separately by different organisations but must be integrated to understand the status of the natural system as it exists today. It will need coordinated efforts, if they have to yield meaningful results.

Now I come to the most important aspect which we forget, i.e., the human being and the ecosystem in which he lives. We tend to look at man in terms of numbers and subject all his problems to statistical analysis. We do not look at man as an individual with a soul but treat him as if he is essentially a stomach to be fed, a body to be clothed and sheltered under a roof. Life is not merely "loading and unloading". We work as if the humanbeing is incidental to all our pursuits of knowledge. We have imprisoned ourselves into impregnable cells in the prison-house atop the ivory tower, far above the madden-ing crowd and its stench and moanings.

What is the ultimate use of all the studies we pursue? It has no relevance unless they are meant to serve man. If so, should not all these studies ultimately be related to man and his needs—present and future? When we look at an area and prepare maps we get an idea of the natural system. The picture that emerges is still "dead" until we bring man into the picture and this is where human geography becomes pivotal in our theme. When we plan development should we plan it only on the basis of natural resources? Should we not consider, how it will affect the human beings. What about the people who live in the area, their traditional talents and skills, and what are our plans to integrate them into the utilization of the resources to generate wealth to help them to live a better life? What about the landless? Can we do something to fit them into a scheme of things when they will not be mere labour, but will live useful lives of fulfilment to themselves and to society at large. In my opinion human geography, and related sciences which study man not in numbers but in terms of their capabilities, are very much neglected. After all, the purpose of all our studies is not merely to produce papers and Ph.Ds, but for the sake of Man. I use the term Man as a being in a space—time continuum, his past, present and future for which all of us, and especially the privileged 'intellectual' elite have a great responsibility.

We belong to a country where everything is considered the visible expression of the Eternal, by whatever name we call it, the invisible thread that runs through dynamism and change. Hence in this land everything—mountains, land, soil and river, plants, trees and animals—have been considered to be sacred and worshipped, notwithstanding our misunderstanding or not understanding the background of these ancient truths or the contradictions that exist between practice and precept. The Red Indian Chief concluded his historic letter thus: "If we sell you our land, love it as we loved, care for it as we have cared for it, hold in your mind the memory of the land as it is when we take it, and with your strength, with all your might, and with all your heart, preserve it for your children and love it as God loves us all. One thing we know, your God is the same God. The Earth is precious to Him. Even the white man cannot escape the common destiny". I repeat none of us, white, or red or black or brown—none of us can escape the common destiny!

The ancient Indian sage declared: "Vasudhaiva Kutumbakam". He did not say merely that the world is one family. The Sanskrit world"Vasudha" means the Earth. He said that the Earth is one Family! What a grand concept! When we hear words of wisdom from modern "sages" who, having come to realise profound truths after having bungled through to their "discovery", let us remember that the ancient Indian sage had said 3000 years ago that the entire earth is one family! The Earth, its hills and mountains, its rivers and streams, its trees and rocks, in fact inanimate and animate, Nature is one Family, is one Integral System!

It must be clear to all that we are on the threshold of the most challenging crisis ever faced by Man, the ecological crisis. The solution of this will need far greater ingenuity than putting man on the moon. It will need something more than knowledge; it will need wisdom. Schumacker said (p. 28-29) : In short we can say today that man is far too clever to be able to survive without wisdom. No one is really working for peace unless he is primarily working for wisdom . . . the exclusion of wisdom from economics, science, and technology was something which we could perhaps get away with for a little while, as long as we are relatively unsuccessful; but now that we have became very successful, the problem of spiritual and moral truth moves into the central position. . . . There can be growth towards a limited objective, but there cannot be unlimited, generalised growth. It is more than likely, as Gandhi said that the "Earth provides enough to satisfy everyman's need, but not for everyman's greed".

In the last paragraph of the book entitled "The Limits to Growth", the authors of these studies carried out in connection with the Club of Rome's Project on the "predicament for mankind" (Meadow's et. al., 1975, p. 197) and have this to say: "The last thought we wish to offer is that man must explore himself—his goals and values—as much as the world he seeks to change. The dedication to both tasks must be unending. The crux of the matter is not only whether the human species will survive, but even more whether it can survive without falling into a state of worthless existence".

The crux of the matter is that the answers to these problems will have to be conceived by individuals with integrity, dedication and a zest for living. We have to realise our highest human potential utilising knowledge and wisdom for the well-being of all. Mirdad declared "Inexhaustible is Man". If we pursue our studies with this vision, "as if people mattered" then we will be able to play our crucial role since society will use our ideas as coming from "people who matter". I conclude with ancient Vedic prayers :

Lokaasamsthah sukhinobhavantu
(May the entire world live in happiness)

* * *

Om sarvavai sukhine santu (May all be happy)
Sarve santu nirasamaya (May all be healthy, free from disease).
Sarve bhadranipasyanthu (May all see auspiciousness)
Maa kaschid dukhamapnuyat (May none suffer)
Om Shantih ! Shantih !! Shantih !!! (May peace prevail !)

SUGGESTED READINGS

Eckholm, Erik P., 1978 : *Losing ground : Environmental Stress and World Food Prospects* : Indian Edition ; Hindustan Publishing Corporation (1) Delhi, p. 223.

Meadows, D.H. & D.L., Jorgen, Randors and Bohrons III, William W., 1975, *The limits to growth*, A Potomac Associate Book U.S.A., p. 205.

Peter, Lawrence, 1977: *Small is beautiful : A study of economics as if People Mattered*, Indian Edition; Radhakrishna, New Delhi, p. 208.

Toffler, Alvin, 1972 : *Future Shock*, Bartam Edition, U.S.A., p. 561.

APPENDIX

LETTER ADDRESSED TO MR. FRANKLIN PEARCE, PRESIDENT, U.S.A., 1855 BY CHIEF, SEATTLE, WASHINGTON

(*Hindu : 29th May, 1976*)

THE GREAT chief in Washington sends word that he wishes to buy our land. The Great Chief also sends us words of friendship and goodwill. This is kind of him since we know that he has little need of our friendship in return. But we will consider your offer, for, we know if we do not do so, the white men may come with guns and take our land.

How can you buy or sell the sky, the warmth of the land ? The idea is strange to us. Yet we do not own the freshness of the air, or the sparkle of water. How can you buy them from us ? Every part of the earth is sacred to my people. Every shining pine needle, every sandy shore, every mist in the dark woods, every clearing, every humming insect is holy memory and experience of my people.

We know that the white man does not understand our ways. One portion of land is the same to him as the next. For, he is a stranger who comes in the night, and takes from the land whatever he needs. The Earth is not his brother, but, his enemy, and when he has conquered it, he moves on . . .

His appetite will devour the earth and leave behind only a desert. The sight of your cities pains the eyes of the redman. But, perhaps it is because the redman is a savage and does not understand.

One thing we know which the white man may one day discover. . . . our God is the same God. You may think that you own him as you wish to own our land. But, you cannot. He is the God of man. And His compassion is equal for the redman and the white. This earth is precious to Him. And to harm the earth is to heap contempt on its Creator. The whites too shall pass—perhaps sooner than the tribes. Continue to contaminate your bed and you will one night, suffocate in your own waste.

When the buffaloes are slaughtered, the wild horses tamed, the sacred corner of the forest heavy with the scent of men, and the view of the ripe hills blotted by talking wires . . . . where is the thicket ? And, what is it to say Goodbye to the shift and the hunt ?. . . . .The end of living and the beginning of dying. There is no quiet place in the white man's cities. . . no place to hear the leaves of Spring, the rustle of insect wings. But perhaps I am a savage and do not understand; the clatter seems only to hurt the ears. And . . . .what is there in life if a man cannot hear the lovely cry of the whippoorwill or the argument of frogs around a pond at night ? The Red Indian prefers the sound of the wind, itself cleansed by the midday rain or scented with a pine. The air is precious to the redman, for, all things share the same breath—the beast, the trees and the man. The white man does not seem to notice the air he breathes. Like a man dying for many days, he is numb to the smell.

We might understand if we know what the white man dreams; what hopes he describes to his children on long winter nights, what visions he burns into their minds, so that they will wish for tomorrow.

But we are savages. The white man's dreams are hidden from us. And, because they are hidden, we will go our own way. If we agree, it will be to secure our Reservation you have promised; there perhaps we may live out our brief days as we wish.

When the last redman has vanished from the earth and the memory is only the shadow of a cloud moving across the prairie, these shores and forests will still hold the spirits of my people, for, they love this Earth as the new-born loves its mother's heart-beat.

If we sell you our land, love it as we loved it, care for it as we have cared for it; hold in your mind the memory of the land as it is when we take it, and, with all your strength, with all your might and with all your heart, preserve it for your children and love it as God loves us all. One thing we know, your God is the same God. The Earth is precious to Him. Even the white man cannot escape from the common Destiny.

TWO

Donald R. Coates

GEOMORPHOLOGY AND OPEN SPACE

INTRODUCTION

A GOOD working definition for open space is provided in 1959 statutes of California (Sec. 1, Chap. 12, Div. 7, Title 1):

> ......any space or area characterized by (1) great natural scenic beauty or (2) whose existing openness, natural condition, or present state of use, if retained, would enhance the present or potential value of abutting or surrounding urban development, or would maintain or enhance the conservation of natural or scenic resources.

Although the Federal Housing Act of 1961 defines open space in somewhat similar terms as undeveloped urban lands that have value for parks and recreation, conservation of land and other natural resources of scenic and historic purposes, there are many dissimilar attitudes and perceptions about such land. Environmentalists view it as sacred whereas business interests consider it land that remains after all other properties have attained their highest and best commercial use. When unused, governments have traditionally viewed it as a waste of potential tax revenue. Open space that has been retained for some societal purpose is termed "deliberate" while land waiting for the developmental process is called "vacant". Open space takes many forms and related terms include *green belt*, *new town*, and *garden city*. The scale of open space ranges from small areas within a planned community to large tracts beyond city and suburban limits. Since this presentation treats open space as a land resource that needs careful management for its preservation, it automatically becomes the domain of the geologist.

HISTORICAL BACKGROUND

The principle of open space dates back at least to the 13th century B.C. when the Levitical cities were surrounded by pasture lands for use by city dwellers. The city of Gezer contained 22 acres with an open land belt 15 times larger. The 1515-16 writings of Sir Thomas More provide green belt ideas where the imagined cities of Utopia were surrounded by agricultural lands; when cities filled new cities were built beyond the agricultural belts. In England the open space concept was rooted in the royal forest and hunting preserves which evolved into public places, "the commons", for gathering of firewood, grazing of animals, and recreation. When faced with increasing pressures from landlords and business to partition the urban commons Parliament passed the Enclosure Act of 1845 which placed restrictions on commons-type land. With the help of the Green Belt Act of 1938 and the Town and Country Act of 1947, the original 74 London commons, each averaging 160 acres, had expanded to 840 square miles in 1959 with concentric rings of agricultural and recreation land mixtures. The Moscow green belt constitutes 30 per cent of the city's total area, and the Ottawa green belt will contain 37,000 acres for farming, forestry, recreation, and water supply protection.

The Garden City concept introduced in 1898 by Ebenezer Howard represents another open space approach. The term "new town" has become associated with those new cities where environmental considerations for open space are planned (Coates, 1974, p. 378-382). Howard's ideas were embodied in the 1903 construction of Letchworth, England, and new towns now exist in many countries as Sweden, the Soviet Union, France, Finland, and West Germany, Columbia, Md. and Rosten, Va. are the best examples in the United States.

The smallest scale for open space retention occurs on the community level and is currently known as PUD—"Planned

Unit Development" (Hantoon, 1971), or open space subdivision (Forster and Kuplesky, 1972). Such developments should not be confused with the subdivision clusters of row house construction, or Levittowns, that mushroomed after World War II. Instead lot sizes under PUD are smaller than conventional zoning code allowance, but the developer agrees to use the residual open space for parks, recreation and scenic sites. Radburn, N.J. built in 1929 served as the prototype for such developments and Heritage Village, Conn. built in 1964 has become the community most mimicked during the past decade. The Town of Ramapo N.Y. provides a larger scale for this design whereby in the process of "density averaging" has obtained 135 acres of public parks.

The Problem

During urbanization local governments have generally translated adjacent lands into taxation domains in terms of industry, commerce residences and public facilities. The basic assumptions were that: (1) economic and population growth are necessary, (2) that unlimited amounts of land are available for development, and (3) that any negative expansion effects are easily solved by human ingenuity. By the 1950s such postulates were finally undergoing critical examination and the country was awakening to a new conservation mood. Land has been disappearing fast and was becoming increasingly costly (Whyte, 1969), pollution was being recognized as an unwanted product of unrestrained growth (Baxter, 1974), and the visual and natural amenities of the landscape were being destroyed by the newly named process of *urban sprawl* (Whyte, 1958). In the past 15 years urban sprawl has been annually consuming one million acres of land, and causing desecration and degradation of 100,000's acres of contiguous lands. Chicago is expanding at a rate of 15.6 square miles per year and paving over important mineral resource areas (Risser and Major, 1967). Woodward-Lundgren & Associates (California Division of Mines and Geology, 1971) calculated that by the year 2000 California will have lost $22.4 billion in mineral production because of urbanization. Indeed the cities of the 1960s are terrain gluttons, using three times as much land as cities of the 1920s; one-third of Los Angeles is devoted to automobile selling, servicing and transportation systems.

Rantz (1970) showed that most of the $62 million flood damages in 1969 from torrential rains in southern California were localized in areas where urban sprawl was not matched with zoning codes for hazard areas. In a similar California study Slosson (1969) determined that of landslide losses in 1969 totalling $6.5 million, only $182,400 damages occurred in areas where geologists had been thoroughly involved in developments.

GEOLOGY INVENTORIES AND PRIORITIES

Geomorphologists are becoming increasingly influential in showing that the quality of the open space environment can no longer be taken for granted, and that reassessment of priorities is necessary to meet the challenge of the new malaise. This new environmental attitude is aimed to show that urbanization disturbs the land, interferes with natural processes, produces excessive long-range damages to renewable, non-renewable and landscape amenity resources, and consumes excessive amounts of the landscape.

Before open space decisions can be properly made inventories must be prepared and priorities established. Open space serves many functions which can be classified into the following units (Williams, 1969) : (1) Resource production in forestry, agriculture, rocks and minerals, water supply and use; (2) Preservation of natural and human resources, fish and wildlife habitats, geologic features, historic and cultural sites; (3) Health, welfare, and well being for water and air quality, waste disposal, recreation, visual amenities; (4) Public safety regarding floods, unstable hillslopes, airplane flight paths, fire zones, (5) Corridors for power transmission and transportation systems, and (6) Urban expansion for commerce, industry, housing, and public service. As shown in numerous studies many of these functions are inter-related and input for their allocation and understanding rests on geologic and earth science-related data. Prior to the 1950s most geologic studies and maps were prepared for use by other geologists. However, starting in the 1950s the U.S. Geological Survey started producing maps and tables of the San Francisco Bay area that were understandable to laymen that described the economic geology and engineering properties of materials. On January 1, 1970 the Survey joined with the Agency Department for Housing and Urban Development in additional inventorying of all geologic features needed in urban planning. More recently the Survey has completed environmental planning maps of the Hartford, Conn. area that show such features as flood-prone areas, soils, landforms, and depth to water characteristics. Other early reports that stressed geologic consideration in land use and open space planning include those by McComas (1968), Mozola (1970), and Christiansen (1970).

Inventories that stress the importance of geologic parameters when land is viewed as a resource in open space planning take two different forms. The first type is characterized by the Metropolitan Washington Council of Governments (1968) inventory of the region for the purpose of determining *environmental constraints to development*. For example shallow bedrock, position of water table, ecologically fragile areas etc. were considerations that determined the type of construction that a location could sustain. The second type inventory has become known as the McHarg approach (or ecologic approach, geologic approach) whereby planning priority is placed on *preservation and protection of natural processes*. Since the early 1960s McHarg and his followers (McHarg, 1969; Wallace, 1970) have placed highest priority on open space and includes all surface water and riparian lands as far back from water bodies as possible. This includes wetlands, flood plains to the 50-year recurrence level, and all aquifer recharge areas. Slopes steeper than 12 degrees should remain undeveloped.

The scale of urban inventories ranges from such small towns as Vestal and Owego, New York (Town of Vestal, 1974); Gentili, 1974) to counties and megalopolis-type regions (Syracuse-Onondaga County, 1973 Nassau Co. Environmental Management Council, 1974). All of New York State has been inventoried (N.Y.S. Office of Planning Coordination, 1971) where land use was divided into 10 main categories, that contain 23 subheadings and 134 categories.

Reports that deal with the establishment of land use priorities for open space range from simple listings of recommendations (Kent, 1970; Gratzer, 1972; Town of Vestal, 1974), to long textual discussions (Platt, 1972), and computerized modelling (Fabos, 1973; Tilmann, Upchurch, and Ryder 1975). McHarg uses a combination of techniques that include a series of map overlays and they can be keyed to a matrix system in which the various environmental components are ranked into numbered classes. Little (1968) shows that when a municipal official is faced with a choice of whether to preserve natural process, community amenity, or recreation that the decision is easy because they are all inter-related. By making natural process the cornerstone of planning other landscape elements are kept in greatest balance. Landslide-prone topography is best retained as open space for water recharge areas. This will also cause less pollution and reduction in sediment yield (Guy, 1970). When aquifer systems receive maximum recharge and streams remain unchannelized river regimen is stabilized and flooding reduced to a minimum (Leopold, 1968). Proper management practices in agricultural land conservation, restriction in flood plain occupancy, and retention of wetlands (U.S. Corps of Engineers, 1974) are additional methods for maintenance of the integrity of open space and provide the maximum techniques for working with natural process systems.

An integral part of open space geologic planning is the location and production of mineral resources. Flawn (1971) discusses some of the attendant problems associated with mining near developed communities. For example aggregate, so necessary in construction industries, such as sand, gravel and rock are not recognized in federal legislation as mineral substances and so are not applicable under the mineral location laws governing the public domain. Although simultaneous multiple land use is possible for subsurface mining, surface materials when mined often create hardships because of nuisances such as noise, dust, vibrations, traffic, and ugliness. Political pressures for their shutdown can become intense. Public education and understanding can aid in developing plans for sequential multiple use and flexible zoning programmes initiated for such resource areas. Thus such lands can be restored to other beneficial uses after the extraction period has ended. Denver provides a worthy example where a sand and gravel pit was mined out, then used as a landfill site until the late 1950s, after which it became the site for the Denver Coliseum.

GOVERNMENTAL ACTION AND LAWS

Once priorities have been established the protection and preservation of open space is accomplished by one or a combination of purchase, zoning restrictions and prohibitions, taxation relief, and other techniques. The California Open Space Act of 1959 provided pathfinding legislation for cities and counties to acquire open space lands and the California Land Conservation Act of 1965 allowed cities to require that subdividers donate park land or make payments in lieu of them. In 1961 Congress passed the Housing Act and started providing financial aid for open space projects, and in 1965 the Bureau of Outdoor Recreation through the Land and Water Conservation Fund initiated a programme for purchase of land if devoted to parks and recreation facilities. Such programmes operate on a 50-50 matching basis, and an example of such usage is the Susquehanna Riverside Project in the Binghamton Metropolitan area where $ 200,000 is being spent for preservation of river banks and public use in the flood plain corridor. During the 1960s Westchester County, New York acquired nearly 2,800 acres of open space lands at a cost of $ 5 million, and the New York Environmental Qualit Bond Act of 1972 authorized $ 175 million out of the $ 1.15 billion total funding, for purchase of lands in the public interest.

A combination of zoning and ordinances have proved effective legal weapons on the local scene to prohibit or restrict development in environmentally sensitive and hazardous areas. Zoning that leads to protection of natural processes is called "conservation zoning" (Kaiser and others, 1974, p. 155-157; Heller, 1974, p. 70-71). Examples of such zoning include flood plain zoning, coastal zoning, wetland zoning, stream bank zoning, and steep-slope zoning. In the latter case management of hillsides in southern California comes under their system of grading ordinances (Jahns, 1969) whereby geologists are deeply involved in planning and monitoring of developments. Other types of zoning that affect open space include large lot zoning, agricultural zoning, impact zoning and multiple use zoning.

Tax relief is becoming an increasingly popular method for preservation of open space. New York's General Municipal Law (1963, Section 247) states that a town can accept less than fee right in open land and in return can provide property tax relief. The usual practice is to reduce property valuation if the owner agrees to leave his lands undeveloped for a stated period of time. The amount of tax is then based on market value of products from the land instead of as real estate operating under the highest and best use principle. California and Pennsylvania passed similar legislation in 1965, and Minnesota followed in 1967.

A variety of other laws have also been passed that affect the management of open space. The Hawaii Land Use Law of 1961 and the Maine Site and Development Law of 1970 are typical statutes that provide constraints to construction. In 1972 Boca Raton, Fla. passed a referendum that established a ceiling growth for construction of new dwellings. Other attempts to inhibit growth include such actions as taken by Governor Tom McCall of Oregon in reduction of the 1972 travels advertising budget and his statements urging visitors not to come to Oregon. In New Mexico a citizens group called the Undevelopment Commission propagandizes against immigration to that state.

The role of the geomorphologist is now firmly established in the analysis of open space and those planning strategies that can most effectively serve to manage this land resource (Alfors, Burnett, and Gay 1973; Legget, 1973; Coates, 1974; Cooke and Doornkamp, 1974; LaFleur, 1974). The special expertise he brings to such evaluations is his ability to predict which practices will cause the greatest disruption of the land-water ecosystem, which lands are most hazardous, and determine the location of natural resources. Thus knowledge of geomorphology, economic geology, engineering geology, and hydrology need to be blended in all appraisal for potential land use changes of the open space environment.

SUGGESTED READINGS

Alfors, J. T.; Burnett, J.L. and Gay, T.E. Jr., 1973; Urban geology master plan for California: Calif. *Div. of Mines and Geology Bull.* 198, p. 112.

Baxter, W.F., 1974: *People or penguins—The case for optimal pollution*: Columbia Univ. Press, p. 110.

Christiansen, E.A., ed., 1970 : Physical environment of Saskatoon, Canada : *Natl. Res. Council Canada Publ.* No. 11378, p. 68.

Coates, D.R., ed., 1974 : *Environmental geomorphology and landscape conservation* : Vol. II. Urban Areas : Dowden, Hutchinson & Ross, Stroudsburg, p. 454.

Cooke, R.U. and Doornkamp, J.C., 1974 : *Geomorphology in environmental management*: Oxford Univ. Press, London, p. 413.

Fabos, J.G., 1973 : Model for landscape resource assessment : Part I of the Metropolitan landscape planning model : Univ. of Mass. *Agri. Exp. Sta. Res. Bull.* No. 602, p. 141.

Flawn, P.T., 1971 : Mineral resources and multiple use land: in *Environmental planning and geology*, D.R. Nichols and C.C. Campbell, (eds.), U.S. Govt. Printing Office, Washington, D.C., p. 22-27.

Foster, J., and Kuplesky, H., 1972 : *The open space subdivision*: New York State Office of Planning Services Albany, p. 28.

Gentili, J., Compiler, 1974 : *Owego environmental study* : Coll. of Architecture, Art and Planning, Cornell Uni., p. 103.

Gratzer, M.A., 1972 : Land use inventory for open space planning in eastern Connecticut : Univ. of Conn. *Agri. Exp. Sta. Res.* Report No. 38, not paginated.

Guy, H.P., 1970 : Sediment problems in urban areas: U.S. *Geol. Survey Circ.* 601-E, p. 8.

Heller, A., (ed.), 1971 : *The California tomorrow plan* : William Kaufmann, Inc., Los Altos, Calif., p. 120.

Huntoon, M.C. Jr., 1971 : *PUD a better way for the suburbs* : The Urban Land Institute, Washington, D.C., p. 71.

Jahns, R.H., 1969 : *Seventeen years of response by the city of Los Angeles to geologic hazards* : Proc. of the Geol. Hazards and Public Problems Conf., U.S. Govt. Printing Office, Washington, p. 283-295.

Kaiser, E.J. et al., 1974 : *Promoting environmental quality through urban planning and controls* : U.S. Environmental Protection Agency, Washington, D.C., p. 441.

Kent, T.J. Jr., 1970 : *Open space for the San Francisco Bay Area*. Inst. of Governmental Studies, Univ. of Calif., Berkeley, p. 85.

LaFleur, R.G., 1974 : Glacial Geology in Rural Land Use Planning and Zoning: in *Glacial Geomorphology*, D.R. Coates, ed., Publications in Geomorphology, State Univ. of New York, Binghamton, pp. 374-388.

Legget, R.F., 1973 : *Cities and geology* : McGraw-Hill Inc., New York, p. 624.

Leopold, L.B., 1968 : Hydrology for urban land planning—A guidebook on the hydrologic effects of urban land use : *U.S. Geol. Survey Circ.* 554, p. 18.

Little, C.E., 1968 : *Challenge of the land* : Pregamon Press, New York, p. 151.

McComas, M.R., 1968 : Geology Related to Land Use in the Hennepin Region, 111. *State Geo. Surv. Circ.* 422, p. 24.

McHarg, I.L., 1969 : *Design with nature* : Natural History Press, Garden City, N.Y., p. 197.

Metropolitan Washington Council of Governments, 1968 : *Natural Features of the Washington Metropolitan Area* : The Council, Washington, D.C., p. 49.

Mozola, A.J., 1970 : Geology for environmental planning Monroe County, Michigan : *Mich. Geol. Surv. Div.* Report Invest. 13, p. 34.

Nassau County Environmental Management Council, 1974 : *Nassau County Environmental Plan Report* : The Council, Mineola, N.Y., var. pages.

Platt, R.H., 1972 : *The open space decision process* : Univ. of Chicago, Dept. of Geog. Res. Paper No. 142, p. 189.

Rantz, S.E., 1970 : Urban sprawl and flooding in southern California : *U.S. Geol. Survey Circ.* 601-B, p. 11.

Risser, H.E. and Major, R.L., 1967 : Urban expansion—An opportunity and a challenge to industrial mineral producers : 111. *State Geol. Survey Environ. Notes* No. 16, p. 19.

Syracuse-Onondaga County Planning Agency, 1973 : *Onondaga County Environmental Plan* : Vol. I. *Environmental Policy Plan*, and Vol. II *Environmental Inventories and Analysis* : The Planning Agency, Syracuse, 173 total pages.

Tilmann, S.E., Upchurch, S.B. and Ryder, G., 1975 : Computer-assisted geologic evaluation for land-use planning : *Geol. Soc. Amer. Bull.*, Vol. 86. No. 1.

Town of Vestal, 1974, The town of Vestal open space inventory : Vestal, N.Y., p. 85.

U.S. Corps of Engineers, 1974 : A community decision, U.S. Army Engineer District, Baltimore, p. 30.

Wallace, D.A., 1970 : *Metropolitan open space and natural process* : Univ. of Pennsylvania Press, Philadelphia, p. 199.

Whyte, W.H. Jr., 1958 : Urban Sprawl : *Fortune*, V. 57, January, var. pages.

Whyte, W.H. Jr., 1959 : A vanishing resource : Open space, now or never: *Landscape Architecture*, Vol. 50, No. 1, pp. 8-13.

Williams, E.A. directed, 1969 : *Open space* : *The choices before California*: Diablo Press, San Francisco, p. 187.

THREE

Donald R. Coates

ENVIRONMENTAL LAW

GENERAL STATEMENT

ENVIRONMENTAL law as a subdiscipline of the legal profession became popular starting in the late 1960s. Although prior to this geologists and lawyers were involved with legal matters that were environment-oriented, there was no formalization with such ruberics in university courses or in law societies. Of course governing bodies have passed many statutes and legislation dealing with the environment throughout recorded history, and the courts have dealt with environmental decisions for millennia.

Law consists of societal rules established by the governing authorities. The regulations that are imposed or mandated have the force of the general public behind them and non-compliance can result in some type of penalty. Laws may be written or unwritten such as English Common Law. They consist of a wide range of legislative actions, ordinances, agency regulations, and codified directives of recognized public officials. Environmental law is now being interpreted to cover an extremely broad spectrum of human endeavours, but this article will emphasize the geology aspects of the subject.

LEGAL ASPECTS OF ENVIRONMENTAL POLICY

Geologists can be involved in a wide range of legislation and litigation through a great variety of different types of environmental law. These laws run the gamut from the general to the specific and their impact on nature and society has been steadily increasing in the past 30 years with exceptional growth in the last 15 years.

General Laws

These are laws that are applicable to many different areas of societal needs and justice systems, but which have been adopted for application into the environmental arena. In the United States the basis for these legalities is founded in The Constitution, especially the Bill of Rights and Amendments, and in the English Common Law. The Fifth and Fourteenth Amendments provide that no person can be deprived of life, liberty, or property without due process of law and that just compensation must be made when property is taken for public use. Such statutes provide the basis for the doctrine of eminent domain. . . . . that governments may acquire private land when in the interests of the public good. The Ninth Amendment is being increasingly used by environmental groups when specific issues arise that do not seem to have been specified in the Constitution. English Common Law has also provided a powerful heritage, and its many aspects have been repeatedly stated in courtroom trials. The combination of the Constitution and English Common Law had produced the prevailing theories for environmental litigation until the 1960s which witnessed the passage of the new type of environmental laws. Thus the usual environmental grievance being debated in the courts involved a plaintiff bringing suit under the claim of damage caused by acts of nuisance, negligence, or trespass.

Nourse v. City of Russelville (1936) is representative of nuisance theory. The city of Russelville, Kentucky in 1935 outlawed cesspools and privy vaults after building a modern sewage treatment plant. The new ordinance was based on the judgment that private disposal units caused contamination of the soil. The plaintiff attempted to enjoin implementation of the ordinance on his land, but the Kentucky court ruled that municipalities may abate nuisance providing it is reasonable.

Specific Environmental Law

Although specific environmental legislation largely dates from the 1960s, two types of environmental laws were enacted through many of the preceding decades. The Rivers and Harbours Act of 1899 was revolutionary in its purpose and represented a model to be followed more than 60 years later with clean air, clean water, and clean land acts. The Act prohibits the dumping or discharging of industrial "refuse" into navigable waters without the permission of the U.S. Army Corps of Engineers. Polluters can be fined up to $ 2500 per day as well as jailed for a period up to one year. Not until the 1960s was this Act enforced, however, at which time many individuals sought to take advantage of the clause whereby a person who provided information that leads to conviction is entitled to one-half the fine as a bounty for discovery. The authority of the Corps to grant such permits was challenged in Zabel and Russell v. U.S. Army Corps of Engineers (1970). The plaintiffs brought suit in 1967 after the Corps had refused them a permit to dredge and fill wetlands in Tampa Bay, Florida for construction of a trailer park. The decision of the appellate court upheld the authority of the Corps which had based their denial on adverse ecological effects that would be caused by the construction.

Specific acts that relate to the environment can be placed into two broad categories : (1) those that mandate user standards, and (2) those that set management policy. The Clean Air Act of 1967 is typical of legislation that provides numerical and quantitative stipulations as to the limits that air can be polluted. By their very nature such acts are highly prone to disagreements in interpretation and thereby result in many court cases. In United States v. Bishop Processing Company (1970) the government brought action against a rendering plant in Maryland because its odours crossed the state line into Delaware. The injunction was upheld by the U.S. Supreme Court to prohibit continuation of such pollutants. The National Environmental Policy Act of 1969 (NEPA) provides another approach for setting standards by environment users. This act can be considered a Bill of Rights for the environment because it stipulates a set of guidelines that must be followed by governmental agencies whenever construction projects are planned that might effect the environment. The environmental impact statement clause in the act paved the way for an entirely different type of appraisal system that must be initiated by the user. It requires that alternative plans must be considered, and that the type and scope of impact on the environment must be analyzed and predicted before construction can commence. Many states have passed similar legislation that enlarges such requirements to state agencies, municipalities, and large private development schemes. The recently passed federal legislation on strip mining (1977) and the many strip mining acts of the states are other examples of laws that mandate quality control by users.

There is an unusually broad spectrum of laws that provides perspectives about the environment and control the management policies by which it is to be governed. Each sector of the environment is covered by a variety of laws and jurisprudence as formulated in court actions in the United States. Many of these have similar stature in other countries of the world. For example in Brazil the Forest Law of 1959 prohibits construction work above specified slope levels in certain parts of the country to reduce landsliding. In an early recognition of the deleterious aspects of urban sprawl England passed the New Towns Act of 1946 which gave the government the authority to designate which land areas were most appropriate for the development of new urban sites. Japan had laws about parks as early as 1873, a forest preserve law in 1907, and a National Park Law of 1931.

Laws for Geologists

Although geologists may be involved in many different types of environmental law, some of this law is aimed entirely at the geologist and requires his participation. In the Raymond Basin Adjudication of 1949 (Pasadena v. Alhambra) the California Supreme Court established that hydrologists were the ones to determine the allocation of water in the region. The grading ordinances in southern California, such as the city of Los Angeles Grading Ordinances of 1952, 1963 and 1969 require that engineering geologists must make the decisions for developments taking place on hillslopes of the city. The Pennsylvania Solid Waste Management Act of 1969 requires geological inspection, analysis, and approval of sites that are to be used as disposal sites for solid refuse. In addition governmental agencies may extend their powers as administrative overseers of environmental affairs into codified regulations for adherence by users. For example in 1972 the New York

State Department of Transportation issued guidelines entitled "Requirements for Geologic Source Reports". Before a mining company can qualify to use products for use on state highways, it must file a special report which can only be undertaken by a certified geologist.

MINING LAW

Laws regarding mining in the western world had their first formulation in the Napoleonic Code of 1804 which held that each landowner possesses everything on and below his property. Nineteenth century monarchies, however, excluded gold and silver which was owned by the crown as a regalian right. In early mining law cases the word "mineral" was interpreted by the courts to mean metalliferous deposits, but by 1854 the Hartwell v. Camman case in New Jersey established that minerals need not contain metal. Additional cases extended minerals to include granite (Armstrong v. Granite Company, 147 N.Y., 495), and limestone (Brady v. Smith, 181 N.Y., 178).

The California Gold Rush of 1849 helped establish new local mining laws that later formed the basis for common law that became incorporated into Federal statutes in 1866 and then into the Mining Act of 1872. Such rules held that (1) discovery and working gave the right to a mineral, and (2) the discovery had the right to continue mining a vein even when it extended under the surface claim of another. Such law was appropriate only for deposits in place, and guidelines were set and gradually enlarged that circumscribed the type of assessment work that was necessary each year to sustain a claim.

The Federal Mineral Leasing Acts of 1917 and 1920 removed coal, petroleum, oil shale, potash, phosphate, and sodium salts from mining controls. The multiple Mineral Development Law of 1954 attempted to resolve problems that had arisen between the mining laws and the leasing laws, and set the stage for the Multiple Use Act of 1955, which stipulated that underground mining was subordinate to surface extraction of most marketable mineral products.

Because of their mobility the oil and natural gas laws have certain differences from the fixed mineral laws. Although the courts have generally held them to be minerals, they cannot be claimed as such until extracted by a landowner and put in his possession, as in Funk v Holdeman (53 Pa. St., 229, 1866).

WATER LAW

Modern water law comprises a mixed blend of statutes, practices, and court decisions from many different countries. Roman law, Moslem law, Spanish law, the Code of Napoleon, and English common law have all provided elements in the matrix of surface and subsurface and legal rights.

Surface Water Law

In the United States the two principal theories that govern ownership rights of rivers are the riparian doctrine and the prior appropriation doctrine. Under special circumstances the principle of prescriptive rights is used.

The riparian doctrine holds that water use accompanies ownership of the land adjacent to the stream. Such a right exists even if the water is not physically used by the landowner. However, when the water is used it must be for reasonable and beneficial use by the owner on his property. Typical lawsuits establishing these principles were in Colburn v. Richards (Mass., 1816), Cook v. Hull (Mass., 1820), and Merritt v. Brinkerhoff (N.Y., 1820). Stroble v. Kerr Salt Co. (N.Y., 1900) set an important precedent whereby the relative economic merits and investments of property owners should not be a consideration in determining rights. The riparian system of water law is especially adaptable for climates where water is abundant, as in eastern United States. However, it is not practical in drier climates where water is scarce and rivers may not flow throughout the year . . . as in much of western United States.

In the more arid parts of the west a water system of rights evolved that was termed the "prior appropriation doctrine". Land ownership is irrelevant to water rights, and instead such rights are acquired by those who first use the water for some beneficial purpose. Thus priority in time is the principal determinant when there is an insufficient amount of water to satisfy all who wish to use it. Lawsuits that established such legality include the California court decision in Tartar v. Spring Creek Water and Mining Company in 1855, and the Desert Land Act of 1877 affirmed such rights to the various states for all non-navigable waters of the public domain.

Groundwater Law

Because less is known of the movement and occurrence of subsurface waters when compared to surface water, laws are in greater conflict when dealing with groundwater. The usual rule established in most courts is to regard percolating waters as part of the property, thus subject to ownership by the land-holder according to the Napoleonic code, as in Acton v. Blundell in 1843. However, if water was judged to flow in "underground streams" the courts ruled it was governed by riparian doctrine, just as surface streams.

With increasing demands for subsurface water in arid regions, other approaches to water law have been made. In the classic case of Pasadena v. Alhambra (Calif., 1949) the court formulated a new doctrine of mutual prescription whereby the original users were given correlative rights, and the total usage of each was to be determined by scientists in the state government.

LAND USE LAW

There are many laws that govern land utilization by man that have been passed on all government levels. Such legislation is aimed at protection of man or for the necessity of providing assurance of his health, safety, welfare, and sustenance.

Water and Soils

Prior to 1849 flood control had been considered only a local problem, but the Swamp Land Acts of 1849 and 1850 recognized the severity of flooding. These acts set the stage for state-wide planning of flood-prone areas and led to the establishment of the Mississippi River Commission in 1879 whose mandated duty was to improve navigation and "prevent destructive floods". A Flood Control Act of 1923 authorized expenditure of $60 million for flood control work on the lower Mississippi but it took the disastrous floods of 1935 and 1936 for Congress to pass the much more inclusive Flood Control Act of 1936. This latter statute along with the Soil Conservation Act of 1935 put the federal government in the business of large-scale land managers. The Department of Agriculture (through the Soil Conservation Service) was to assume responsibility for upstream flood control measures, and the Army Corps of Engineers was authorized to undertake projects on navigable rivers throughout the United States. The scope of the Department of Agriculture was enlarged in 1954 with passage of the Watershed Protection and Flood Prevention Act which allowed this agency to construct flood prevention structures that would impound up to 4,000 acre-feet of water if costs did not exceed $250,000.

Another type of breakthrough in environmental land law occurred in 1902 with the Reclamation Act which established irrigation in the West as a national policy for implementation. Although lawsuits were brought against the government challenging such powers, the Supreme Court affirmed the principle of governmental promotion for the general welfare through "... large-scale projects of reclamation, irrigation, or other internal improvement" (United States v. Gerlach Livestock Co., 339 U.S. 723, 738, 1950). An unusual expansion of governmental prerogatives for the general welfare occurred in 1933 with enactment of the Tennessee Valley Authority Act which charged the new agency with a land development and rehabilitation programme that would involve land use, water control, and soil conservation throughout the region.

Range and Forest Lands

The first federal environmental legislation of forested lands was passed in 1891 with the Forest Reserve Act, and followed in 1897 with the Forest Management Act. These acts were aimed at the West and their purpose was to insure a continuous timber supply and provide favourable conditions for streamflow. The Weeks Act of 1911 laid the groundwork for federal programmes on a country-wide basis, and the Clarke-McNary Act of 1924, the McSweeney-McNary Act of 1928, and the Sustained Yield Forest Management Act of 1944 combined to provide a comprehensive environmental package of the legal stewardship of the government of forest lands. The 1944 Act summarized this posture by stating their purpose was ". . . . in the maintenance of water supply, regulation of stream flow, prevention of soil erosion, amelioration of climate, and preservation of wildlife". Again in 1960 the Multiple Use-Sustained Yield Act emphasized the necessity of maintaining forests in prepetuity for the purpose of assuring: (1) timber from trees, (2) forage from grazing lands, (3) recreation, (4) wildlife management, and (5) protection of rivers and consideration for their constant flow.

The Taylor Grazing Act of 1934 was the first federal legislation that directed attention to a public policy for the establishment of grazing districts and the raising of forage crops. It provided the Secretary of Interior with powers to undertake erosion and flood control measures that will protect and rehabilitate grazing lands.

Parks and Wilderness

Governmental attitudes for the establishment of park-like areas were greatly influenced by Roman law and the English "commons" which held that certain lands be set aside for the people as a public trust. Passage of the Yellowstone Act of 1872 by Congress made Yellowstone National Park the first of its kind in the world, and put the government into the business of preserving terrain for recreational purposes. The New York legislature enacted a bill in 1885 to establish a forest area that was to remain permanently ". . . as wild forest lands" and led to the Adirondack State Park. This "forever wild" concept culminated in the Wilderness Act of 1964 which set aside millions of acres throughout the United States for the purpose of benefiting ". . . . . the American people . . . an enduring resource of wilderness".

Additional environmental legislation was implemented by such federal laws as the Land and Water Conservation Act of 1965 and the Wild and Scenic Rivers Act of 1968. These bills amplified the government's resolve to plan and preserve lands for outdoor recreation and ". . . . enjoyment of present and future generations".

LEGAL RIGHTS FOR NATURAL FEATURES

It is becoming increasingly common that people, the courts,

and legislative bodies are recognizing the importance of maintaining balance in natural systems. Since ecosystems cannot speak or act for themselves in legal proceedings various surrogate organizations are becoming their advocates and are championing the view that natural features and objects have certain legal rights. The large number of lawsuits, as well as their significance, that challenged unnecessary and environmentally harmful developments were instrumental in passage of NEPA and establishment of the Environmental Protection Agency. Furthermore the states, such as Hawaii, Vermont, and Maine were passing land use laws whereby the land was to be treated as a natural resource. The dual rationale for these trends was that man no longer was to be considered as omnipotent, and that destruction of nature ultimately created feedback mechanisms that proved harmful to man as well.

Perhaps tempering with rivers and floodway corridors received the earliest attention and prohibitions of construction. More recently coastal areas have been selected as sites for preservation. Typical of these laws is the Delaware Coastal Zone Act of 1971 to " . . . protect the natural environment of its bay and coastal areas . . ." The Coastal Zone Management Act of 1972 was the first land control legislation ever passed by Congress, and it has led to much additional legislation by states that border coastal areas. Wetlands are another environmental feature that are being increasingly protected by law. Massachusetts and Connecticut were some of the earliest states to pass such legislation, and the Freshwater Wetlands Act of 1975 in New York is one of the most comprehensive of its type for the protection of these features.

EMINENT DOMAIN

Governments have always used their power to acquire private lands in the interest of the public good. Such lands in the United States cannot be condemned for whimsical or nonbeneficial uses, and the rights of property owners to obtain just compensation are assured by the Constitution. However, numerous environmental lawsuits result from condemnation practices when:

(1) Plaintiffs believe abnormal environmental damage will occur. For examples see Coates (1973, p. 383-387) in discussion of such cases as Hetch Hetchy in California, Boundary Waters Canoe Area in Minnesota, Mineral King in California, and the Florida jetport hearings.

(2) Plaintiffs feel aggrieved because of the lowness of the monetary assessment of their property, (see Coates, 1971, and 1976).

RECENT ENVIRONMENTAL TRENDS

In addition to the laws previously mentioned, and closely allied statutes, a new range of legal strategems have been employed during the past 20 years in attempts to cope more effectively with environmental problems (see also articles in this volume on Geology and Open Space, Environmental Management and Urban Geomorphology). The largest number of these laws is designed as protection for lands and resources adjacent to urban areas. The following list provides a sampling of the legal techniques that are used to minimize environmental degradation and to reduce property and financial losses to the public.

(1) Limits to growth. To preserve land and prevent urban sprawl communities, such as Boca Raton, Florida, are passing various types of statutes that place ceilings on new housing and developments.

(2) Preservation of farm land. To prevent the loss of valuable farm land governments ease the tax burden on farmers by assessing property at values much less than "highest and best use" of the land. Incentives are also provided to the farmer, such as in New York's Agriculture District Law of 1971, to retain his lands in productive use. Suffolk County, New York, through this act is aiding the farmer cause by subsidies of millions of dollars.

(3) Preferred site locations. By various ordinances and zoning laws governments prevent the exploitation of some lands and allow for the development of others. One method being used is illustrated by the Flood Disaster Protection Act of 1973 whereby property owners and communities cannot qualify for insurance coverage unless they meet location and structural guidelines stipulated under the law. This act

provides for losses incurred by floods or mudslides if the appropriate insurance measures have been consummated.

Thus, environmental law plays an exceptionally important role in the lives of people and the well-being of society. The topic is now a recognized area of specialization in law (Sax, 1970; Landau and Rheingold, 1971; Murphy, 1971) and is gathering momentum each year. Legislation and precedents set in court cases, cover the entire spectrum of man's relation to the environment. Geologists are playing an increasingly significant part in environmental legal affairs because problems of land, pollution, energy, and resources generally fall within his area of expertise.

SUGGESTED READINGS

Coates, D.R., 1971: Legal and environmental case studies in applied geomorphology: in *Environmental Geomorphology*, D.R. Coates, ed., Publ. in Geomorphology, State Univ. New York, Binghamton, pp. 223-242.

Coates, D.R. (ed.) 1973 : *Environmental Geomorphology and Landscape Conservation*: Volume III. Non-Urban Regions: Dowden, Hutchinson & Ross, Stroudsburg, p. 483.

Coates, D.R. 1976: Geomorphology in legal affairs of the Binghamton, New York, metropolitan area: *in Urban Geomorphology*, D.R. Coates, ed., Geol. Soc. America Special Paper 174, p. 111-148.

Kiersch, G.A. and Cleaves, A.B., eds., 1969: Legal Aspects of Geology in engineering practice: *Geol. Soc. America*, Engineering geology case histories No. 7, p. 112.

Landau, N.J. and Rheingold, P.D., 1971: *The Environmental Law Handbook*: Ballantine Books, New York, p. 496.

Murphy, E.F., 1971: *Man and His Environment: Law* Harper & Row, Publishers, New York, p. 168.

Sax, J.L., 1970: *Defending the Environment*: Vintage Books, New York, p. 252.

FOUR

D. K. Ghosh

GEOMORPHOLOGY OF RAVI BASIN IN NORTHWEST HIMACHAL HIMALAYAS AND ENVIRONMENTAL MANAGEMENT

INTRODUCTION

A SYSTEMATIC geomorphic study has been attempted for terrain classification and their significance with the aid of aerial photos and surface characters in the districts of Chamba and Kangra within Ravi basin bounded by Lat. 32° 14′ and 32° 48′ and Long. 75° 50′ and 76° 32′ covering parts of toposheet numbers 43P and 52D. Presently, the knowledge of geomorphology of the region is very sketchy and hence an appraisal of terrain types, drainage basin, river valleys and the morphometric study to understand the history of geomorphic evolution in this part of the Himalayan belt has been brought out to assist in the study of basin management. The gamut of geomorphic description of Ravi basin in the region initially dictates the need for understanding the geologic events reflecting the relief and hence the paper highlights first the rock description and structural set-up and then the present day landforms along with their influence on basin management.

SUMMARISED GEOLOGIC EVENT

The mountain blocks in parts of Chamba and Kangra districts within the Ravi basin are composed of a series of differing architectural elements represented by sedimentary, metamorphosed sediments and igneous massifs (Fig. 4.1) in the following tectonic sequence from southwest to the northeast.

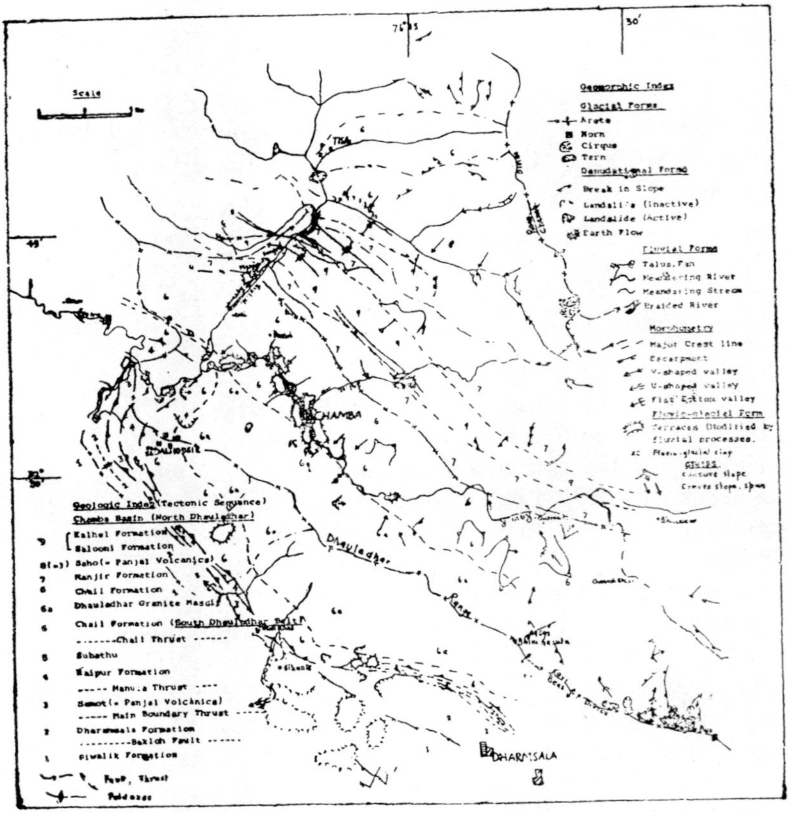

FIG. 4.1 : *Geomorphic and Geologic Map of a part of Northwest Himachal Himalayas in the Districts of Chamba and Kangra, H.P.*

North-East	*Chamba Basin*
Kalhel Formation	Limestone, dolomite and quartzite.
Salooni Formation	Carbonaceous slate, phyllite and interbedded thin limestone.
Saho (=Panjal) Volcanics	Compact and amygdaloidal basalt.
Manjir Formation	Pink or grey limestone at top, pebbly slate, laminated slate.
Chail Formation	Slate, phyllite and some schist, silty quartzite.

Dhauladhar Belt

Dhauladhar Granite Massif

Chail Formation	Slate, phyllite, quartzite and mylonite gneiss.

Chail Thrust

Subathu	Shale and limestone.
Raipur Formation	Slate, phyllite and limestone.
Mundi Formation	Dolomite and lokhan bed.

Manuta Thrust

Samot (=Panjal) Volcanics	Altered basalt, tuff and intertrappean shale.

Main Boundary Thrust

Dharamsala Formation	Sandstone and shale.

Bakloh Fault

Siwalik Formation	Sandstone and shale.

South-West

The rock units have uniform northeasterly dip of varying magnitude and have been thrown into isoclinal folded sequence. Whereas the north Dhauladhar belt shows an open asymmetric synclinorium with the main axis trending in NW-SE near

Kalhel and veering to NE-SW across Siul; the south Dhauladhar belt comprises three tectonic slices. The area has witnessed major tectonic uplift due to compression and igneous intrusion [Fig. 4.1(a)]. The shallow Archaean basin is considered to have experienced compression with the deposition of older sediments (now metasediments) during Puranas, thus forming a synclinorium designated as Purana Geosyncline. Lines of fracture already existed in the basement which facilitated igneous intrusions. During late pre-Cambrian, further compression of basement floor and uprise of granite magma initiated the formation of a sub-marine ridge separating the major basin into Chamba and Dalhousie basins with a deeper basin to the north and a shallower one to the south of submarine ridge.

Then followed a break in sedimentation from late pre-Cambrian to lower Carboniferous. Again a major tectonic uplift occurred in the area due to volcanic eruption during upper Carboniferous. The lower-middle Tertiary was the last phase of igneous activity causing sinking of the basin to the south and uplift to the north. The entire process of sedimentation, combined with uplift and exhumation has formed inliers of older metamorphics and folded outliers of the younger rocks. Faulting appears to have taken place along the axes of the tight anticlinal flexures due to compression and uplift. The overall relief forms thus represent the impact of folded rock in a geosynclinal basin in which anticlines have been eroded and uplifted blocks have been subjected to vigorous dissection. Due to structural complexity, it has not been possible to precisely indicate the possible planation surfaces and old residual relief forms. Uplifts during pre-Cambrian, upper carboniferous, lower-mid Miocene and early pleistocene might have developed planation surfaces, of which Chamba basin surfaces were raised more than south of Dhauladhar ones. Tectonic vigour, causing deformation, is a source of heterogeneity in the landscape of Chamba-Kangra region and thus it proliferates difference over short distance.

LANDFORMS

Two types of landforms—erosional and depositional have been recognised in the area (Fig. 4.1). Both are contemporaneous

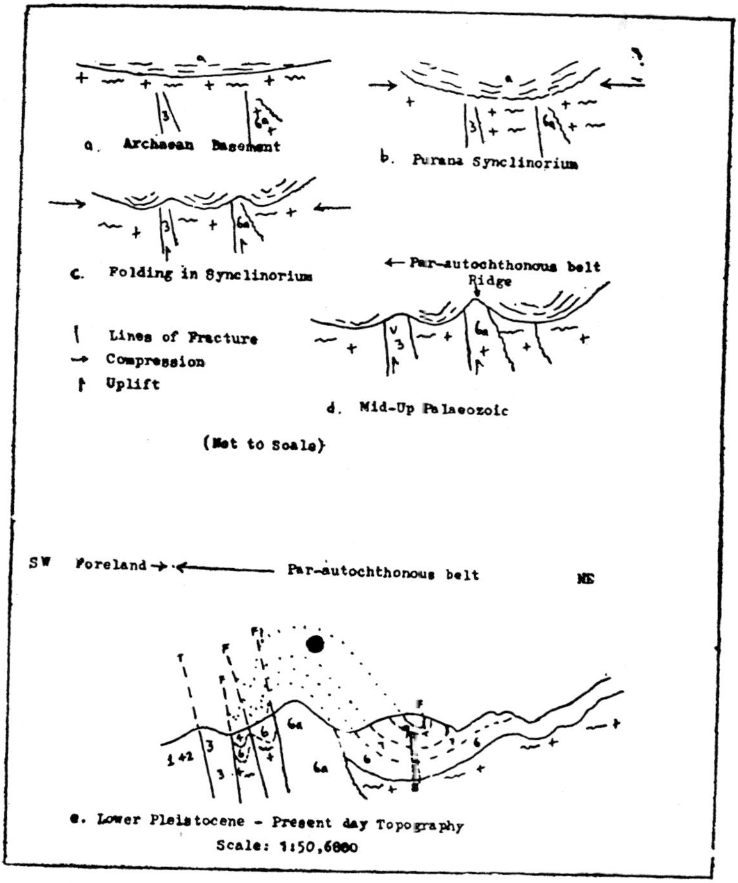

FIG. 4.1 (*a*) : *Sketch showing stages of Tectonic Evolution Around Chamba-Kangra Region, H.P.*

Note: Sedimentation in the basin continued up to Triassic accompanied by compression and uplift.
For Index please see Fig. 4.1.

with the former preceding the latter. These include three types of geomorphic features—glacial, denudational and fluvio-glacial, which have later been modified by fluvial processes and these have been identified on the basis of the presence or absence of U-shaped and V-shaped valleys, aretes, morphometry and other landform characters. The making of the present day landform and landscape is attributed to sub-aerial processes.

Glacial Landform

In the strictest sense, the glacial landforms are generally lacking in the present day period along Dhauladhar Range and on the hills, south of Sach Pass Range. The once glacial forms are being modified by ice during winter months and fluvial action during other periods of the year. A few glaciers, however, exist in the higher elevations around and above 4500 m, near Bhim Gasutri and Singher Pass along Beas-Ravi divide of Dhauladhar Range and near Barakanda north of Bharmaur besides many others along Ravi-Chenab divide. Sufficient quantity of rock debris moved by glaciers in the geological past along the northern slopes of Dhauladhar Range (north of Dalhousie and south of Chhatrari) and along southern slope of Ravi-Chenab divide north of Bharmaur are now in the process of being eroded by present day snowfall and stream action following snow melting. These have been deposited in the glacial lake, known as tarn, occurring as linear depression in the cirque itself or break in slope areas seen along north of Dhauladhar Range and southern slope of Ravi-Chenab divide. The geomorphic expression of the rock debris, designated as glacial moraine and earth flow, is very conspicuous above elevation 4000 m. The shape of the cirque, as ovate depression, has been found to deepen with three sided steep walls in some areas along Dhauladhar Range. The author believes that rotational slip might have assisted in the formation of cirque in these areas (Fig. 4.2). Longitudinal section of the slope consisting of a basin (B) and riegel (R) indicate a rotational movement and this can also be called an erstwhile 'glacial stairway or step'. The author also relates this to valley constriction and resultant ice deepening (causing glacial load), varying rock resistance and preglacial irregularities and fracture spacing. Later, some of the cirques at lower altitude have taken the place of 'tarn'. The Khajiar lake just north of Dalhousie is one such example [Fig. 4.2 (a)]. Post

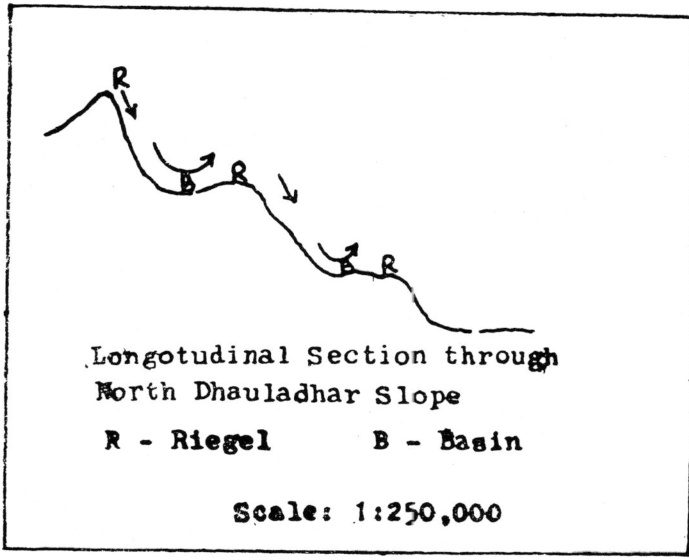

FIG. 4.2: *Rotational Slip Due to Glacial Movement, Dhauladhar Range near Dalhousie and Dharamsala.*

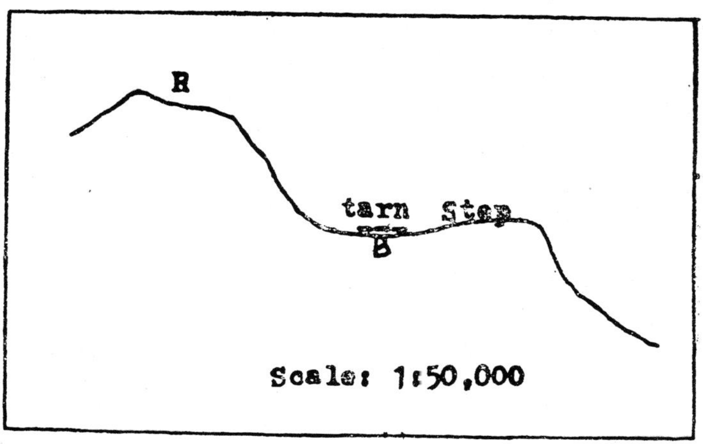

FIG. 4.2 (*a*): *Glacial Step around Khajiar Lake near Dalhousie*

glacial humid climate has transformed the topography to a large extent now by extensive erosion.

Denudational Landform

Masswasting by fluvial or ice erosion is the principal source for the development of denudational forms in the entire area within Chamba-Kangra districts, of which fluvial action prevails in the lower elevation areas below 4000 m. In the dry cirque areas and nearby slopes, north of Dhauladhar Range, ground moraine known as 'earth flow' is conspicuously developed. It shows concentric ring and constitutes finer sediments as a result of rock weathering and is mostly surcharged with water due to melting of ice. Lateral and downcutting by Ravi and its tributaries in the higher reaches of Dhauladhar Range have left a number of semicircular erosional scars due to downslip of earthmass at lower levels of the aforesaid valleys (Plate 1). Besides, it has also been noted that moderate to steep valley slopes, rocks exposed on steeper-than-dip slopes, open relief joints and cross fractures or shears sloping outwardly into the valley are some of the other factors generating landforms as a result of slips and mass movements. Such features are conspicuously seen along the Siul Seawa and Sal valleys forming major tributaries to Ravi, which has comparatively less pronounced mass movement features, probably due to wide valley section. A study of valley cross-section for Ravi from near its source to the mouth of the basin under reference reveals that the watershed divide between Ravi and Chenab has shifted substantially towards north (Fig. 4.3) and such a feature is attributed to active slope erosion in the softer rocks by the tributaries to Ravi. For a case study, a morphological map has been prepared (Fig. 4.4) for the planning Baira-Siul hydel Project in Chamba district (Ghosh, 80), with respect to alignment of open channel along Baira and Baledh valley slopes. The indurated rocky terrain along Dhauladhar Range in the south has stood like a land relief after Cambrian period.

Weathering has also formed a number of geomorphic features along Dhauladhar Range, near Dalhousie. These are fluvio-glacial moraines and till consisting of clays and silts, with big erratic boulders of granite randomly disposed on gentle and break in slope areas. Banikhet, near Dalhousie is one such example for the said deposits.

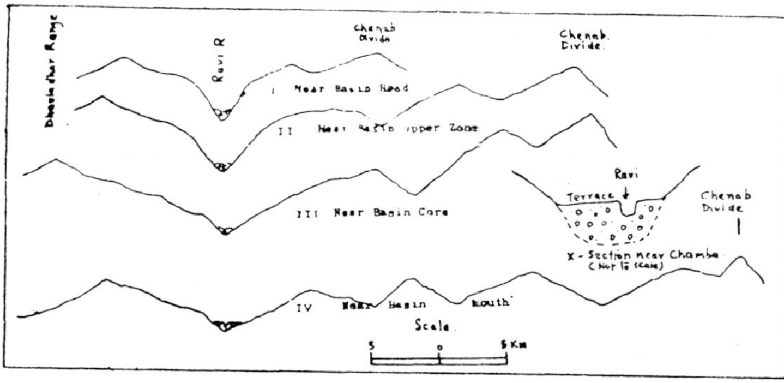

FIG. 4.3 : *Stages of Ravi Valley Development, Chamba and Kangra Region, H.P. (Showing geomorphic sequence and relief development).*

Fluvial Landform

Depositional landforms by fluvial action like talus, fan, terrace and older, recent flood plain deposits are conspicuously seen along Ravi, Siul and Sal valleys and on the peidment slopes in the north and south of Dhauladhar Range respectively. The major terraces are near Manjir and Chamba along Siul and Ravi valleys and Sihunta, Chuari-Khas and Tundi-Khas on the south of Dhauladhar Range (Plate 2). The terraces are found on the banks and break in slope areas and show an admixture of fine and coarse material that has been reworked by fluvial processes. Some of the terraces like the Manjir terrace along Siul valley and Sihunta terrace near Kangra have dissected appearance. Along Ravi near Chamba, two level old terraces have been seen besides a recent flood plain at bed level. Talus and fans (Plate 3) are located at the mouth of major streams like Siul and tributaries and Ravi and tributaries, around Koti and Chamba respectively.

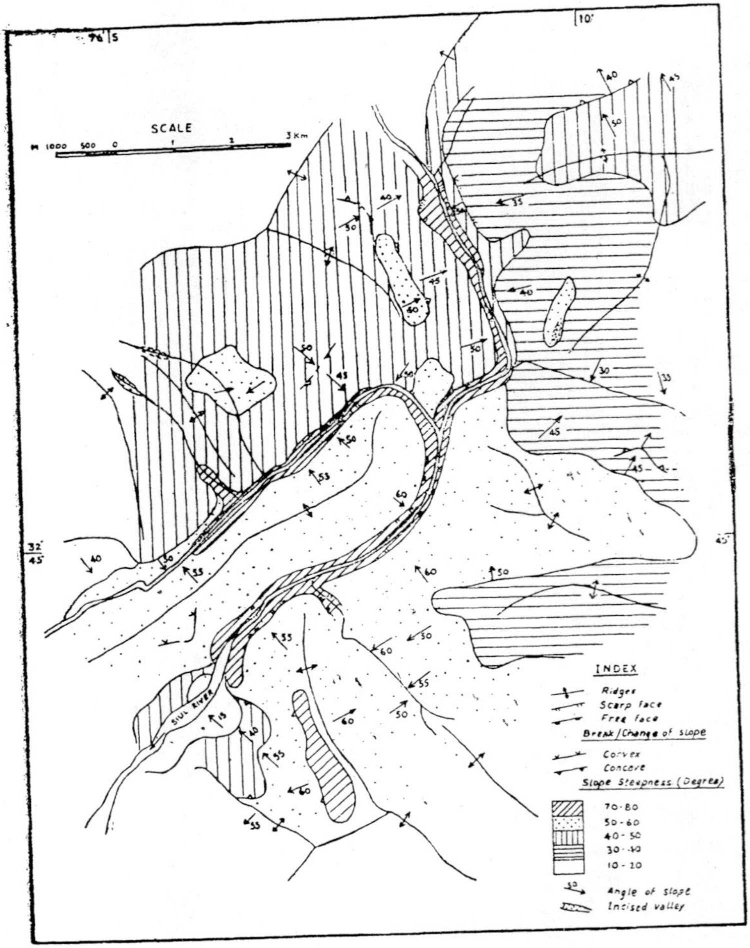

FIG. 4.4 : *Generalised Morphological Map of a part of Siul Sub-Basin, H.P.*

Morphometry

Obsequent hill slopes and near vertical valley slopes are seen around northeast of Sungani in Siul valley, Kuther and Bani in Seawa valley, Holi and Urna in Ravi valley. These have been designated as 'escarpments'. The height of escarpment is approximately 200 m and it runs for at least 0.5 to 1.5 km. Resistant rock like quartzite and granite and some limestone exhibit this kind of feature. Apart from this, velley form as U-shaped in the snowy areas along Dhauladhar Range near Singhar pass and V-shaped in the denudational and fluvio-glacial hills along Dhauladhar Range south of Bharmaur are also seen. Shape and dimension of such valley forms are time dependent and related to bed rock character like easy erosion and inherent fracture.

Regardless of the nature of formation, the steep slopes and escarpments might be controlled by differential movement of two structural compartments (separated by joints or fractures) and this has created fragmentation of bed rock in the Ravi and Siul valleys thereby favouring erosion, scree and slips besides retreat of the escarpments on the southern slopes of Ravi-Chenab divide (Fig. 4.3). In the Siul valley, parallel gullies or streams have developed with steep longitudinal profile in the resistant quartzite and cemented pebbly slate. On the other hand, the softer rock-slate and phyllite in the Ravi valley around Chamba having low escarpments, dissected into narrow ridges, are influenced by active slips.

DRAINAGE SYSTEM

The Ravi and its tributaries like Siul, Seawa and Sal provide the main drainage of the region to the north of Dhauladhar Range (Fig. 4.5). The Ravi flows in an almost SE-NW direction for a larger stretch and then suddenly swings to southward to the north-west of Dalhousie. The Ravi drainage basin is near ellipsoid in shape, (barring minor diversions along periphery) and is elongated in a SE-NW direction upto the point of confluence with Seawa, which is the mouth of the basin for the purpose of present study. The basin is bounded by Beas-Ravi divide in the south and Ravi-Chenab divide in the north. Southern part of the besin i.e. north of Dalhousie, has no other major sub-basin. However, in north, northwest

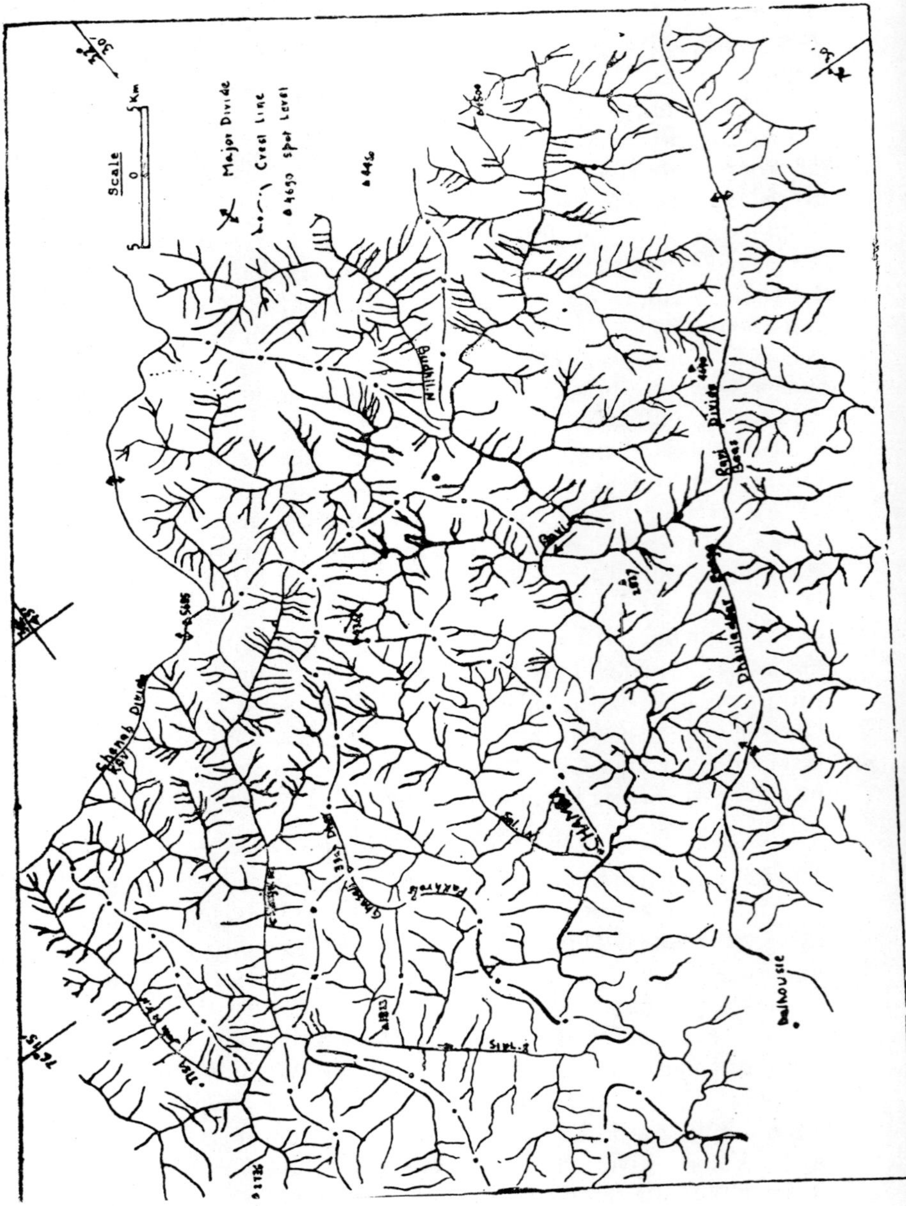
Scale
5 0 5 Km
Major Divide
Crest Line
4690 Spot Level
Chenab Ravi Divide
Budhil N.
Ravi
Ravi Beas Divide
Dhauladhar Range
Chamba
Dalhousie

1. Showing mass wasting features along south Dhauladhar Range

2. Fluvial-glacial terraces along south Dhauladhar Peidmont slope

(a) Chuari terrace (T)

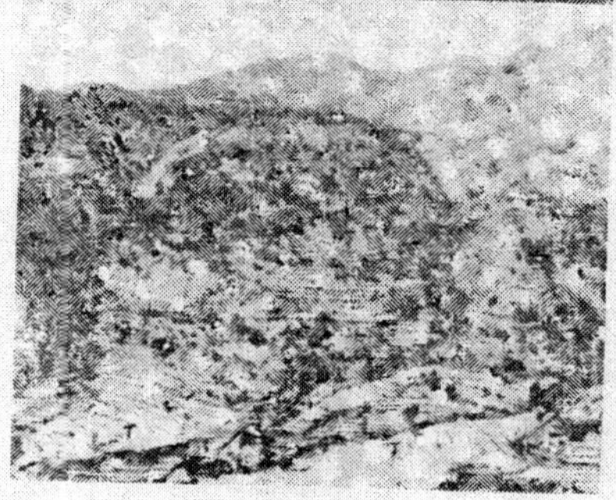

(b) Sihunta terrace (T)

3. Fan terrace along Siul valley near Sungani

4. Panoramic view of Dhauladhar Range

(a) Structural high in sedimentary rock

(b) Sharp crest in granite massif

5. Sculptured topography in upper Siwalik sandstone

6. Topographic high in Siwalik sandstone

and east of Chamba, there are three major sub-basins drained by Siul and its tributaries, Sal and Budhil nalas. Pakhroto-Ghaselidhar is another major sub-divide separating the Siul and its tributaries from the Sal and other sub-basins north of Chamba.

The main drainage system is characterised by initial straight courses for long stretches of the rivers of streams implying structural control. In general, the drainage pattern is sub-dendritic and sometimes sub-parallel with local annular or radial drainage around folded or igneous rocks of Dhauladhar Range.

Development

The geographic location, topographic heights, average rainfall and the prevailing temperate climate, governing humid and vegetated condition, suitably fit in with the prevailing conditions of Dhauladhar Range and its northern terrain. Hence the evolution of drainage phenomenon in the Ravi basin is primarily due to humid condition to which gravity is the instrumental agency for movement of material and surface runoff. The interplay between the draining water and the surface on which it is flowing, because of friction, has developed two types of drainage configuration—one conforming to the structural trends of the bed rock and the other developing on the unconsolidated cover mass, the former being commonly discernible in the area. The initial slope in the higher reaches of north Dhauladhar Range and south Ravi-Chenab divide slope determined the flow of direction (Fig. 4.6). Such areas have only short first order streams. Following Horton (1945), the drainage development system has been worked out. It is noticed in these areas that the system of gullies and channels form on an inclined granular surface i.e. poorly consolidated mass covering the bed rock [Fig. 4.6 & 4.6 (a)]. Later these first order streams deepen and unite to develop an incipient drainage net, with the trunk stream to form a dendritic drainage system so commonly seen in the humid regions. In some rocky areas of Dhauladhar range, occupied by crystallines, the rills and gullies have formed on steep barren slopes directly due to fluvial erosion. The initial valley shape due to

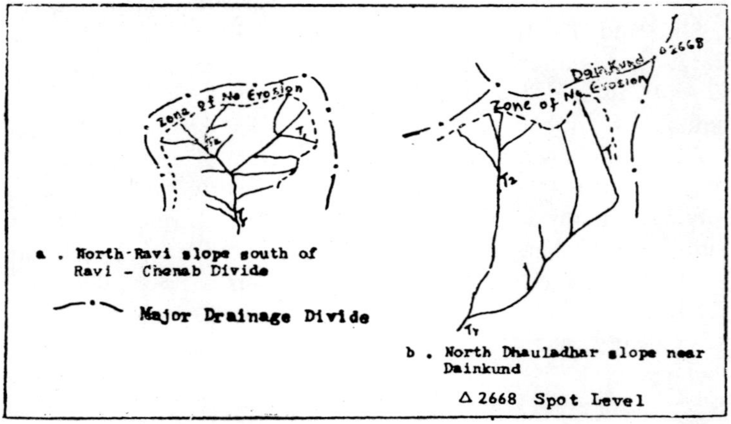

FIG. 4.6 : *Incipient Drainage Net with Trunk Stream around Basin Periphery.*

(T_1—1st order stream. T_2—2nd order stream). *Tr*—Trunk stream

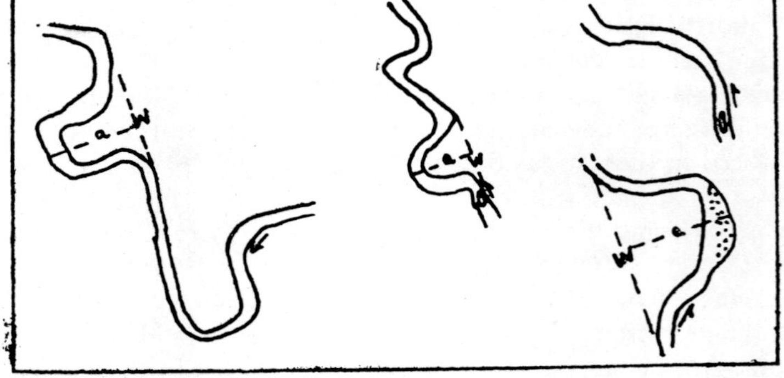

FIG. 4.6 (*a*) : *Geometric parameters of Ravi River Channel*

w—wavelength, a—amplitude, w—width,

glacial action later reworked by fluvial processes and mass wasting, is clearly evident from the valley profile (Fig. 4.3).

Drainage Basin

Ravi constitutes the main drainage basin in the area. A study of the morphometric parameters such as drainage network, basin geometry and intensity of dissection (Table 4.1) demonstrates that correlations exist between: (i) basin area and total stream length, lengths of streams of each order and mean length of streams of each order; (ii) total number of streams and the number of streams of each order; and (iii) stream frequency or drainage density and basin relief.

The study of stream network is based on aerial photos on scale 1 : 64000 and toposheet on scale 1 : 250000. Ravi, forming the trunk stream, has the highest 5th order. The drainage development is approximately 30 per cent on the north Dhauladhar slope i.e. south of the Ravi basin and 70 per cent on the south Ravi-Chenab divide slope i.e. north of Ravi basin and comparing its nature and density in the sub-basins of these two slopes, it is seen that the stream order increases with basin size and the drainage development is up to 3rd order on the north Dhauladhar slope while it is up to 4th order on the south Ravi-Chenab slope. The overall rating of drainage density within 30 streams per 10 sq. km. indicates a low to medium texture in the Ravi basin. Further, it is also noted that the basin has an average of 7 streams of 20 m length per 100 m increase in basin relief. However, the stream length of each order significantly decreases and the development of 1st order stream around 90 per cent decreases appreciably with stream order. No definite relationship could be established between bifurcation ratio and other stream parameters within the Ravi basin, however, the study in smaller sub-basins within and outside Ravi basin has shown a bifurcation ratio of 4.1 to 4.5 and this suggests dynamic equilibrium in the region indicating erosion control.

In general, the drainage pattern is sub-parallel to sub-dendritic with local radial or annular around igneous rock along Dhauladhar range. The coarse texture along higher Himalayas above 3500 m around Dhauladhar and Ravi-Chenab divides and medium texture in the lower reaches, their pattern in different lithounits (Ghosh and Chibber, 76 and 80) have shown that the streams flowing over various rock units exhibit different kinds of channel development. In the tributary streams, the valley shape is predominantly V-shaped with fairly steep bed gradient and the 1st order streams are shorter than 2nd and 3rd order streams by about 50 per cent.

Following Schumen (1963), it is seen that Ravi has a near straight to transitional pattern in the initial 50 km. length upto longitude 76° 44′ and then it becomes transitional to semi-irregular in the next 20 km. length upto long. 76° 30′ and finally in the entire basin it has an irregular meandering course with amplitude axes trending in N-S to NNE-SSW directions. The meander course in Ravi has a varying wavelength and amplitude ranging from 250 m to 2.5 km. and 200 m to 750 m respectively. These meanders in the geological time scale is considered unstable but for river planning, in which the time scale is short, it may be taken as stable and no alteration of meander is possible.

Ravi has a concave channel profile with knick points coinciding with rock outcrops in the upper reaches of the basin. The channel width varies from 75 m to 250 m in the entire basin. This has shown a fair relation with valley depth (as proved at few sites by drilling) and the width-depth ratio is around 0.6 to 1.0. A high width-depth ratio say in Siul valley around Baira confluence and in upper reaches of Ravi near Chatrari has shown large bed load (Ghosh, 68; Ghosh and Kumar, 67). Following Mannings roughness coefficient (n), the river Ravi carries varied size of material having gravel to boulder in large extent. There is greater variation of particle size across channel than along long profiles. Coarse particle is in the deeper portion and fine on the sides. Following Hack (1957), the channel slope has been calculated as 18 m/km. which fairly tallies with the observed data from toposheet.

In the mountain reach of the Ravi basin under study, no relation could be established between channel slope and annual discharge because of varying rock character and channel depth. However, a general assessment indicates that the discharge is more in a straight course than meander course and the wide valley with shallow depth channel is associated with high proportion of bed load, while a deep channel is related to sediment load. As a result, Siul and associated tributaries in upper reaches have shown deep valley fill sediment (more than 30 m in depth). Following Patterson (1970), the discharge in the Ravi basin can be calculated when basin morphometric parameters are known fairly well now and have been given in the preceding chapters.

TABLE 4.1: **Morphometric Variables of Ravi Basin**

Variable with symbol	*Details*
Drainage Network	
Stream Order (u)	1st order to 5th order.
Number of stream of each order (Nu)	1st order (880), 2nd order (139), 3rd order (30), 4th order (4) and 5th order (1).
Total number of streams within basin (Enu).	1054 nos.
Bifurcation Ratio Rb=Nu/Nu+1	3.9 to 5 (approx and wide variation)
Total length of stream of each order (Lu).	1st—1655.5 km, 2nd—533.5 km, 3rd—437.5 km, 4th—123 km and 5th—14 km.
Mean length of stream of each order (Lu=Lu/Nu)	1st—1.9 km, 2nd—3.9 km, 3rd—14 km, 4th—30 km and 5th—14 km.
Total stream length within basin (ELu).	2763.5 km.
Stream length ratio (Ri=Lu/Lu−1)	
Basin Geometry	
Area of basin (Au)	3375 sq. km. (rounded off)
Length of basin (Lb)	75.5 km.
Width of basin (Br)	45.2 km.
Basin Perimeter (P)	4400 km. (rounded off).
Basin Circularity (Rc=Au/Area of circle having same P)	3375/2291=1.5 (rounded off to nearest fraction)
Basin Elongation (Re=Diameter of circle having same P/Lb).	54/75=0.72.

Variable with symbol	*Details*
Intensity of Dissection	
Drainage density (Du=ELu/Au)	2763/3375=09 km/sq.km
Constant of channel maintenance (C=1/Du).	1/0.9=1.1
Stream frequency (Fu=Nu/Au) Fu=Enu/Au	1st—880/3375=0.26, 2nd—139/3375 =0.04, 3rd—30/3375=0.009, 4th—0.001, 5th—0.0003. Total=1054/3375=0.3/sq. km.
Texture ratio (Tu=Nu/Pu)	—
Heights	
Channel slope of Ravi	20m/km=1.15° (approx)
Min valley side slope	25°
Max valley side slope	90°
Height of basin mouth (z)	559 m.
Height of highest point on watershed (Z).	5004 m.
Total basin relief (H=Z−z)	4445 m.
Local relative relief of valley side (h).	15 m to 200 m.
Relief Ratio (Rh=H/Lb)	59.3
Rugged Number (Rn=DxH/5280)	—

TERRAIN CLASSIFICATION

The terrain, occupied by acid and basic igneous rock, the metamorphic and sedimentary units, presents a unique pattern of topography, soil and vegetation cover and such a character has been described as a landsystem of its own. The geomorphic description of each landsystem is given hereinafter:

Coarse Clastic Landsystem of Upper Siwalik (Kutori-Kangra Belt)

Area —South of Dhauladhar Range in a NW-SE belt between north Dunera and Kangra.

Geology —Upper Tertiary soft sandstone and conglomerate.

Geo morphology —Less dissection to ravineous tract or high dissection, coarse drainage of dendritic type, short discordant spurs or ridges of 15-20 m height, bare slope to thin clayey soil slope, wide alluvial terraces up to 1.5 km. in length, narrow and steep valleys at head reaches, large terraces.

Fine to Medium Clastic Landsystem of Lower-Mid Tertiary (Bakloh-Dharamsala belt)

Area —South of Dhauladhar Range in a NW-SE belt along Bakloh-Dharamsala, and some Eocene outliers within pre-Tertiary units.

Geology —Sandstone, claystone and limestone.

Geomorphology —Moderate dissection, low to high relief up to 31 m. long trendline of ridges with wide crests up to 100 m, coarse and sub-parallel to sub-dendritic drainage, steep slope up to 20-25 m in height occupied by resistant sandstone and limestone, low slope upto 10° in soft shale, slate has good land use, transverse streams have gentle bed slope like Chakkikhad. Limestone has erosion features.

Fine to Medium Clastic Landsystem of Permian (Sungani-Kalhel belt) and unclassified South Dhauladhar belt.

Area —North of Dhauladhar Range in the Ravi and Siul valleys in NW-SE belt (Sungani, Kalhel-Bharmaur) and Sherpur-Dhundiara belt in south Dhauladhar.

Geology —Quartzite, limestone, shale.

Geomorphology —Moderate to high dissection, low to mostly high relief upto 35 m, long trendline of crest with narrow to moderate width upto 50 m, sub-parallel to sub-dendritic drainage, steep slope is shown by quartzite and limestone, low slope of about 10-15° in shale has good land use, steep bed gradient in streams, karstic feature in limestone, besides linear ridges.

Metamorphic Landsystem of South and North Dhauladhar Range (Chail Belt)

Area —Bathri-Banikhet-Chuari Khas belt in south Dhauladhar and Pukhri belt in north Dhauladhar.

Geology —Slate, phyllite, schist and quartzite.

Geomorphology —Smooth discordant ridges, erased topography, short spurs, moderate slope upto 35°, except some steep slope in quartzite near Koti, sub-dendritic drainage, good land use in steps, large terraces along rivers, big slumps and mass wasting features.

Basic Igneous Landsystem of South and North Dhauladhar Range

Area —Simbleu-Samot belt in South Dhauladhar and Kalhel-Saho belt in Ravi basin of north Dhauladhar.

Geology —Altered basalt, tuff and intertrappean shale.

Geomorphology —Parallel or branching accordant ridges up to a height of approx. 15-30 m, short spurs, but generally broad undulating terrain, dense sub-dendritic drainage with pinnate tributary pattern, terraced flood plains along large rivers or streams like Ravi, Sal, and Bhedkhad etc. prevalence of small scale slumping, intermittent minor streams and large perennial streams.

Acid Igneous Landsystem of Dhauladhar Range

Area	—Dainkund-Kharugala belt in NW-SE direction.
Geology	—Granite-gneiss of fine to medium grained and porphyrictic variety.
Geomorphology	—Moderate to high relief massive hills, long spurs, wide crests, dendritic to local radial drainage, high slopes covered with thin clayey soil or weathered bed rock, moderate slope up to 60° shows poor land use.

Topographic Elements

These land systems fit in well with the five physiographic belts of the region (Figs. 4.7 and 4.8) and their distribution and disposition have a relation to morpho-tectonic and palaeogeographic history of the region. The Dhauladhar rises from 400 m in the south to approximately 5500 m above mean sea level (Plate 4), and it forms the main watershed and dividing line in the south. In the northeastern part of Ravi basin, the other ranges like Pakhroto-Ghaseli and Ravi-Chenab divides also follow in a nearly WNW-ESE to NW-SE direction corresponding to the Dhauladhar while on the western part the ranges veer to westerly direction. The trend of these ranges by and large correspond mainly with the regional trend of the formations. The Dhauladhar-Ravi basin landscape exhibits an erosional surface and includes flood plains of the main river valleys at lower levels. The erosion surface comprises valley head concave slope, spur end slope being convex in plan and valley side slope which are both straight or curved. It is to be noted that the slope in the area varies in general from 25° to 90° and the relative relief from 15 m to 200 m. The resistant rocks like

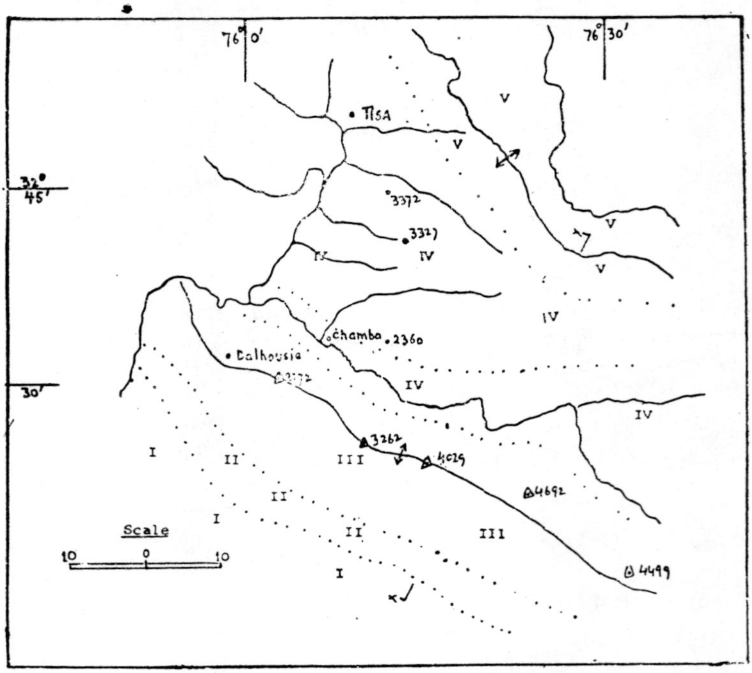

FIG. 4.7 : *Physiographic and Morphotectonic Belts around Chamba-Kangra Region.*

the granite-gneiss of Dhauladhar Range, quartzite and dolomite of Siul valley and basics of Simbleu-Sudal belt of south Dhauladhar present steep convex slope of 60° and above with high relief while the weak rocks like the slate, phyllite and schist of Chamba-Pukhri belt display moderate to gentle slope of 40° to 10° with open concavity. Such a feature extends for long distances. The softer upper Siwaliks present a bad land topography (Plate 5). It can be stated that the prevailing topograpy of the region has peculiar set-up in the folded geosynclinal belt. The granite massif dominated by vertical movement in the geosyncline shows large wedges, structural compartments and vigour of dissection, with break in slope at the contact of uplift and downwarp i.e. granite-Chail

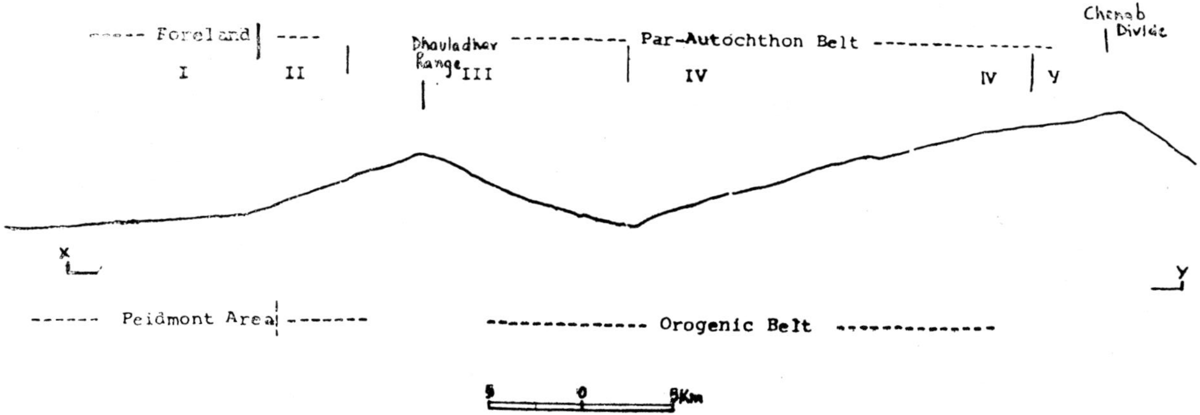

FIG. 4.8 : *Showing Physiographic and Morphotectonic Belts across Beas-Ravi-Chenab Divides*

I—Sub-Himalayas (Siwalik and Equivalent Belt)
II—Lesser Himalayas (Pre-Tertiary Rock)
III—Dhauladhar Range (Crystalline Rock)
IV—Ravi Depression and Dissected Terrain (Metasedimentary and Sedimentary Rock)
V—High Mountain of Ravi-Chenab Divide

phyllite-schist contact. A landscape of crest, narrow valley and long steep slope along top of Dhauladhar range is developed in severally crushed granite rock, and/or fragmented material. A series of talus scree along north Dhauladhar Range is an evidence of valley fracture now occupied by Ravi River. Sinusities of Ravi are however considered to be due to entrenched meanders rather than faults guiding the angularities. A number of elbow bends between the Ravi-Budhil nala confluence in the upper reaches down to the Ravi-Seawa confluence reflect structural control joints. The overall deep valleys in the Ravi basin are indications of relief of geosyncline, characterised by relief inversion along Siul valley, near Dhar and Kalhel [Fig. 1(a)]. There, the topographic anomaly comprises synclinal hills on structural lows. Since the area is intersected by fault as well, the interaction of fold and fault has given rise to perched syncline, dominated by isolated relief forms giving rise to sharp ridges and escarpments in dolomite and quartzite. The anticlinal portion got eroded fast resulting in deep valley section. Such a feature has resulted in the development of valley assymetry in the Siul drainage basin. The Dhauladhar terrain in the south of Ravi occupied by pre-Tertiary rocks has been found to be lacking in structural landform-bed rock relationship and this can be attributed to be independent of structural control, the topographic highs coincide with structural highs, and this feature along Dhauladhar Range might be either due to an intrusion along faulted anticline or an uplift of the basin floor. The sedimentary belt in the south Dhauladhar presents a landform controlled by lithology and structure where topographic high has been seen in the sandstone of Siwalik belt folded into a anticlined (Plate 6).

It is seen that valley side slope forms an important component in the relief system, and hence an attempt has been made to provide a functional sub-division of slopes in the Ravi basin (Fig. 4.9) in which the following five types of slopes have been identified :

(i) *Nose* : It forms the dry area above elevation 4000 m around the basin rim with convex contours, as spur and has runoff

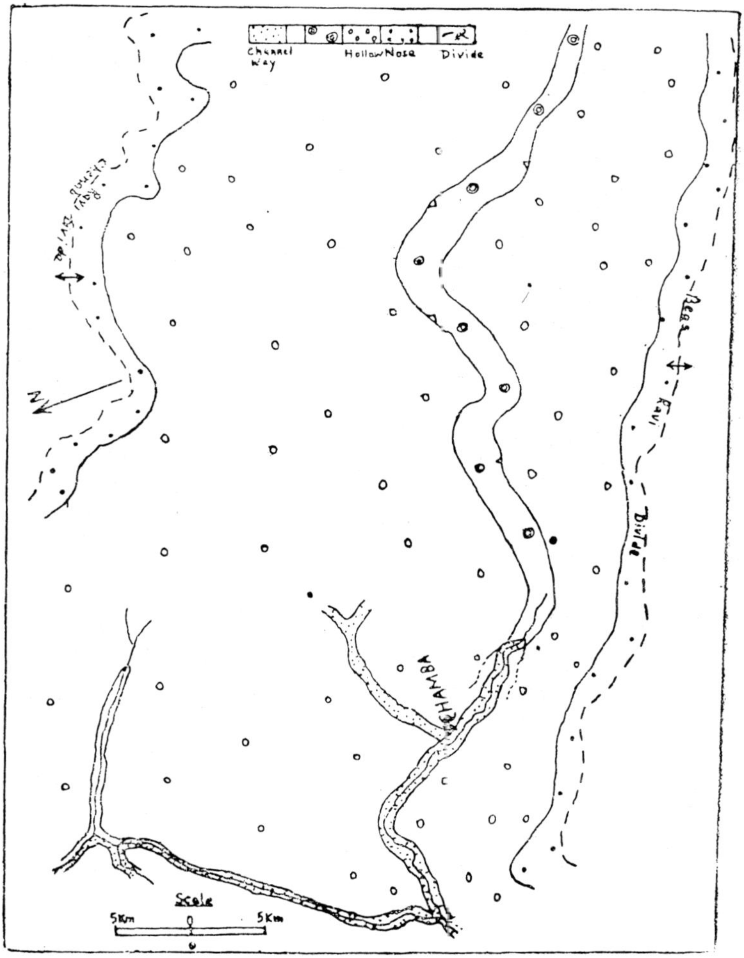

FIG. 4.9 : *Landscape Map of a part of Ravi Basin showing functional Sub-division of Slope, Chamba-Kangra Region.*

proportional to radius of contour curvature.

(ii) *Side Slope* : It has straight contours with appreciable slope length.

(iii) *Hollow* : It has concave contour in the central part of the valley with appreciable runoff.

(iv) *Foot Slope* : It forms the lower part of the side slope.

(v) *Channel Way* : It forms the river course and flood plain areas.

On the basis of this study, a six unit model of the valley side slope, combining the geomorphic processes, has been presented (Fig. 4.10) to characterise the areas suitable for

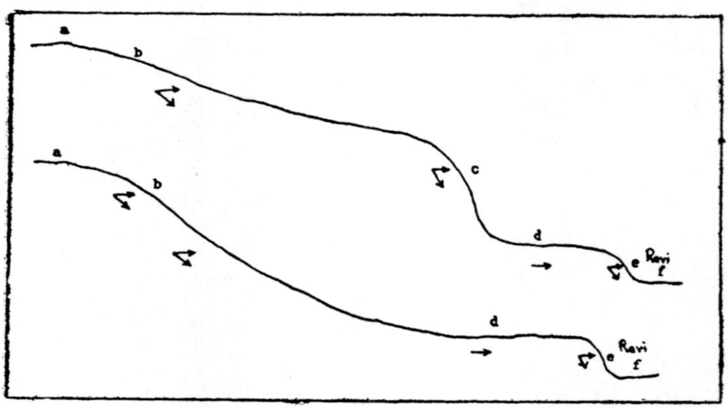

FIG. 4.10 : *Unit Model of the Components of Valley side slope in parts of Ravi Basin, Chamba-Kangra Region.*

a=Areas of infiltration

b=Areas of Soil Creep and mass movement

c=Areas of physical weathering

d=Deposition areas

e=Corrosion areas

f=Aggregation areas

land use and slip potentiality. It has been seen that the areas covering part of 'hollow and channel way' form good land use zones while the other areas are susceptible to slips. The study of typical slopes in the entire Ravi valley (numbering about 80) indicates that weathering and valley incision are actively related to the evolution of slope and the 'slope decline' is the predominant element to form the present day ground surface having convex and concave profile. In the dissected terrain of Dhauladhar and Ravi basin landscape, a number of accordent summits have been noted. These are prevalent in the metabasics and the granite massif. Such accordant summits mark old local planation surfaces which have been modified by dissection in the present day time. However, it is rather difficult to present a map of planation surfaces with the available data but four such surfaces cannot be ruled out.

ENVIRONMENTAL MANAGEMENT

The role of geomorphology in the environmental management of Ravi basin mainly conceives three facets of the terrain—its relief and material, water resources and soil erosion. An explanation of the landform evolution given in the preceding chapters has laid down a basis for this study with a view to analysing their implications on the basin development with respect to rapid urban growth and river valley projects. Accelerated erosion of the landform due to natural processes (like river bank erosion and slope instability); and secondly human activity have already shown their adverse effects in the development of Siul velley (Ghosh and Kumar, 1967 and Ghosh, 1980) due to uneconomical depth of valley fill and regolith slope. Hence an attempt has been made to reassess the developmental activity of Ravi basin in the light of available geomorphic parameters.

The once glacial region is presently humid, and the zone of no erosion around the basin periphery above elevation 4000 m represents a dry atmospheric area. This is perhaps due to uplift of the mountain in the region which has an indirect influence on the discharge of water from the basin. Consequently, the drainage basin (precipitation catchment area) would be depressed below the drainage divide and thus shall reduce runoff from the rising mountain range. Hence the critical runoff flow distance may have to be measured from the lower limit of the dry zone.

Two contrasting mass wasting situations exist in the present day climax humid region of the Ravi basin. Normally there is an adjustment between regolith slope (due to weathering and/or creep) and channel incision. The slopes are steep but stable unless disturbed. The regolith cover is thick and the vegetation root does not appear to anchor the debris with bedrock. High water and clay content are the characteristic features of the regolith slope and these induce instability because of excess rainfall or artificial excavation. Again, the outside banks of bends along water course of Ravi are the sites of accelerated erosion by undercutting and caving. Such features can be observed at the foot of terraces in the vicinity of Chamba.

The predominant meandering habit of Ravi in the region represents a type of dynamic equilibrium (least work) flow form by the presence of longer channel distance and low path of slope in the valley flat basides resulting progradation of channel bank and increased areas of usable land for agriculture. In developing minor irrigation and micro hydel schemes in the region the straightening of the meander might initiate changes in the width, depth and shape of the valley due to accelerated erosion. The author feels that this may have to be encouraged to deepen and enlarge the channel cross-section.

Floods do not normally occur in the Ravi valley as the habitation land is on upslope or high level near flat old terraces or river flood plains which are the relict forms left by glacial effects. Master plan of Ravi basin contemplating a series of dams (Ghosh, 1968) might even outflow levels and thus reduce peak discharge. Apart from this, the proposed dams in the upper reaches meant for controlling deposition of sediment load, might accentuate erosional agency in the middle and lower reaches of the basin. The proposed storage structures in the middle and lower reaches of the basin under study i.e. in the meandering zone, might further reduce bank erosion inhibiting sediment load along Ravi but the tributary erosion may be accentuated due to mouth deepening. Thus any proposal for dams downstream of Chandot Khas (west of long. 76°30′) may lose storage and flood control capacities due to anticipated erosional tendencies by spill release in the upper reaches instead of deposition. In effects low diversion structures are ideally suited in the Ravi basin for irrigation, water supply and power generation.

The discharge measurements, until such time gauge stations are fixed, may include the catchment below the dry zone which does not contribute any runoff. Snow melting in the dry zone of Ravi basin is limited and does not appear to modify the discharge appreciably.

Urbanisation, especially the expansion of Chamba and adjoining cities along Ravi and tributary basins and associated road building activity have induced land erosion and consequent changes in the sediment load and hydrological character by way of lowering of water table due to downward shifting of ground configuration. Runoff and storm flow increase in tributary streams and disposal of untreated waste in Sal and associated nalas around Chamba and in Siul valley colony sites of Baira-Siul project further add environmental pollution.

The free draining flood plains, fan and river terraces, having low retentive capacity, are not good sources for ground water development on sustained basis. Since these are located on the stream banks with a break in slope towards hill face, a low ground water profile from the hill face to the river bed is obtained in the media. This is a major constraint in the terrace landform of the basin, which is otherwise a good area for agriculture and urbanisation. Hence perennial sources provide permanent water supply for local irrigation and urban needs. In and around Chamba and other major urban areas, the perennial streams need to be free from pollution effects. The storm flow and other wastes can well be disposed of in the steep gradient drainages downstream of the urban areas after channellisation, through catch water pits and drains. The cross divide drains are the best sources.

The limestone of the basin is cavernous and presents linear ridges on the surface. The cavities are joint controlled and pose a constant problem for leakage (Ghosh, 1967). In view of this the earlier proposal for a 121 m high across Siul near Salooni had to be abandoned on account of loss of storage.

Communication is another important aspect for development of the basin. The experience from the existing major roads like Chamba-Tisa and Chamba-Sharmaur, which pass through active and inactive slip areas, suggests a need for incorporating geomorphic play of the region in planning road communication. Hence the map showing potential mass wasting features and escarpments can be a guide for such planning and in this situation, the near head of slip and down break of slope areas are the best zones for road alignment. The location of bridge sites is also required to be planned with respect to prevailing mass wasting features.

The presence of fluvial and fluvio-glacial landforms indicates about the character of material suitable for various engineering structures as coarse and fine aggregate. The sand from fan terraces is ideal for cement mortar while the Ravi bed provides more construction material of gravel and boulder size.

CONCLUSION

The mountain belt of the Chamba-Kangra region in the Ravi basin presents a landscape influenced by uplift, exhumation and denudation. The relief forms represent the impact of folded and faulted blocks in geosynclinal basin in which anticlinal crests have been eroded most and the uplifted blocks have been subjected to vigorous dissection. The drainage pattern and stream geometry, mass wasting features due to accelerated erosion and unstable slopes, and the landforms dictate a need for utilising the geomorphic parameters in the environmental management of the Ravi basin.

SUGGESTED READINGS

Cooke, R.U. and Doornkamp, J.C, 1977 : *Geomorphology in environmental management: An introduction.* Clarendon Press, Oxford.

Garner, H.F., 1974 : *The origin of landscape—A Synthesis of Geomorphology,* Oxford University, New York.

Ghosh, D.K. and Kumar, S., 1966 : Progress report on the planning stage geological investigation for Siul Hydel Project, H.P. (Unpublished GSI report).

Ghosh, D.K., 1967 : *Application of Geohydrology to the investigation of Siul Hydel Project, H.P* (Presented at ISEG, Lucknow and Roorkee).

Ghosh, D.K. and Chibber, I.B (1976), Photo-geological and geomorphological report of a part of H.P. Himalayas, Distt. Chamba and Kangra, H.P. (Unpublished IPI report).

Ghosh, D.K., 1980: Aid of photointerpretation in the identification of geomorphic and geologic features around Chamba-Dharamsala area, H.P., (Submitted for IPI seminar, 80).

Ghosh, D.K., 1980: *Geological Influence of Landslips and other mass movements in the planning of Baira-Baledh Siul Link Project, H.P.* (ISL-1980).

Ghosh, D.K., 1968, Geological report on the Ravi Master Plan Projects, H.P. (Unpublished GSI report).

Ghosh, D.K. and Kumar, S. 1967 : Progress, report on the planning stage geological investigation for Baira-Baledh-Siul Project, H.P. (Unpublished GSI report).

FIVE

K. K. L. Das and *K. N. Das*

ALLUVIAL MORPHOLOGY OF THE NORTH BIHAR PLAIN—A STUDY IN APPLIED GEOMORPHOLOGY

INTRODUCTION

THE MORPHOLOGICAL characteristics of the North Bihar Plain, lying wholly to the north of the Ganga river, are exhibited in its alluvial landscape. The fluvial action, river bank erosion, congestion and stagnation of beds, anastomosing channels duly separated by shoals and bars etc. are a function of hydraulic variables to which the rivers of the plain are generally vulnerable. In turn, these give rise to processes of alluviation and diluviation, some of which accentuate flood hazards particularly during peak monsoon storms. Therefore a study in alluvial morphology of the plain is full of absorbing interest. In the present paper, an attempt has been made to study the geological setting, drainage characteristics, physiographic personality of the region together with their 'restrictive' and 'permissive' relationships to agriculture, irrigation, transport and communication, human activities, etc.

GEOLOGICAL SETTING

With the exception of the Himalayan foothills corresponding to the Siwalik strata exposed as the northwest bulge of the Sumeshwar and Dun ranges in west Champaran, almost the entire North Bihar Plain is an area of great physiographic homogeneity throughout its length and breadth from Sidhaw in the west to Thakurganj in the east. All along it forms part of a vast depositional plain of the Ganga drainage system. With the exception in the extreme northwest where the topographic expression imparted to the landscape is anything but plain and homogeneous, the formation of the rugged relief is a mixture of shales, sandstones and clays, etc. The entire area is a remarkably featureless plain composed of river alluvia worn out of the northern mountain rim and deposited in layers over the centuries. This has resulted from the aggradational action of many big and small perennial Himalayan affluents that debouch onto the Bihar Plain. Naturally, this has remained the playground of rivers, e.g., the Ghagra, Gandak, Burhi Gandak, Baghmati, Kosi[1] etc., in addition to minor streams of the Adhwara group, viz. the Kamla-Balan, Tiljuga, Bhutahi Balan, and a host of minor trickles which make their way to the Ganga in meandering, braided and frequently changing courses. The vast alluvial plain of the Ganga system separating the gigantic Himalayas from the Peninsular tableland occupies a depression formed during the last phase of the Himalayan Orogeny. This trough of lowland is estimated to be between 5,000 and 6,000 metres deep.[2] The vast depression, though originally small, was later enlarged due to the sinking floor of the original depression under pressure from thousands of metres of sand, silt and clays brought down by the rivers issuing forth from north to south. This aggradational plain stretches roughly over 51,800 sq. km. Between the Himalayan foothills in the north and course of the Ganga in the south, the entire North Bihar Plain is covered by river alluvia and is composed of silt, sand and clay in varying proportions from one section to another. There are ample evidences that river-borne alluvium was gradually carried down by young and active northern channels in course of thousands of years and concomitantly laid down in a long and shallow estuarine gulf that existed over the surface during the Tertiary era.[3] The thickness, character and general relief of the plain owe much to the infilling of this gulf.

The physical landscape of the region carries the records of successive phases and processes of its evolution. Since the last phase of the Himalayan Orogeny, i.e. Pleistocene, the whole region has undergone sedimentation which is amply justified by thick deposit of alluvia lying conformably over the basal rocks of old geological periods. It is only in the northwestern extremity in West Champaran district where the sedimentary rocks of the Siwalik system lie exposed, folded and uplifted above the present level of alluvium. The Dun and Sumeshwar hills are a continuation of the Siwalik range and are formed of ill-compact and premature sandstones and clays. Probably, these Himalayan foothills were originally anticlinal 'the southern part of the anticline having disappeared as there are no fossils in these strata'.[4] On the lower slopes of the hills, gravel beds are found covered with *sal* dominated thick forest trees. Near marshy conditions exist in the terai region where dense forest is generally found.

The rest of this vast land is almost flat in character and the geological structure is by and large simple. It consists mainly of *bhangar* or old alluvium and *Khadar*, i.e., new alluvium. Both of these constitute as alluvial deposits mainly of sand, clay, silt, gravels and coarse fragments. The latter is composed of silt, clay and sand and is renewed every year by fresh deposit brought down by active streams, e.g., the Ganga, Gandak, Burhi Gandak, Baghmati, Kamla-Balan and the Kosi which engage themselves in fluvial action. As a matter of fact, relatively new deposits mainly occupy the southern parts of the region where the streams converge onto and meet the Ganga river. Normally, these sediments are surely of fluviatile origin consisting of sand through loam to alkaline clay as in the Gandak basin to the west whereas in the intercones of the Gandak and Kosi usually fine texture is met with. On the other hand, old alluvium is found farther away from river courses, especially on uplands of the plain where silting is a rare phenomenon. A slight uplift of the plain on the whole resulting from continual deposition by a host of perennial rivers has imparted variable character to the plain that will be discussed later. This type of alluvium is found mostly in northern parts of Purnea, Saharsa, Madhubani, Sitamarhi and parent Champaran districts which are close to *terai* region of North Bihar and which generally lie beyond the reach of actively flowing streams. One of the main features of the old alluvium is the occurrence of nodular concretions of carbonate of lime, known as *Kankar* that are found here and there.

DRAINAGE CHARACTERISTICS

The study region abounds in rivers, streams and small

drainage courses which are mostly Himalayan affluents and are, in cases of big rivers, perennial. These Himalayan streams such as the Gandak, Burhi Gandak, Ghaghara, Baghmati and its extension Kareh, Kamla-Balan, Bhutahi Balan, Tiljuga, Kosi and the Mahananda, etc., are major streams. They are responsible for the upbuilding of this plain. All these streams having north-south or northwest-southeast trends flow nearly parallel to each other, and on the completion of their journey they empty into the Ganga. A host of minor streams together with a few others including the dead and defunct *dhars* carry flood waters during the monsoon rain (June-October). The Ganga river flows from west to east along the southern margin of the region and forms the regional boundary.

The morphology of the plain is, to a large extent, dependent on drainage characteristics. In course of long journey flowing down through mighty Himalayan ranges and then the *terai* zone of Nepal, these streams experience a sudden break in slope and become obliged to carve out braided and meandering channels on getting a level surface downstream. This occurs also due to the nature of the bed material. In order to have free flow of flood waters through a monotonous flat surface they often change their courses. While braiding is more prominent feature along the Kosi in the east, meandering courses are by far the most noticeable feature in the central and southern parts. Braided pattern is found in the Kosi-ravaged belt of the Saharsa district. Soon after the Kosi enters the plain below Chatra the river begins to flow through many divided channels of braided pattern. They bifurcate through interlacing courses at one place and reunite at the other. For a distance of nearly 210 km. from the southern end of the Kosi gorge near Chatra (Nepal) to as far south as Sonbarsa, the Kosi river is wholly braided.[5] The bed of the river is nearly two km. wide with islands and bars, embayments and depressions alternating with each other throughout its longitudinal section. Meandering courses and migration of channels constitute probably an important characteristic of the plain. The study area is representative of a dendritic drainage pattern. In their upper section, tributaries towards the Ganga look like branches of a tree converging toward the master river. Due to change in gradient most of these Himalayan affluents form inland deltas with apex in the north and alluvial fan spreading out toward the Ganga in the south. All these features give rise to deposition that become a boon as a solid base for agricultural activities that are carried on in the fertile plain. Economic activities, however, coexist with flood hazards of very serious proportions.

Frequent changes in courses of rivers constitute another important characteristic feature of fluvial geomorphology. In the Adhwara floodplain, old courses of the Kamla show that they have changed their way several times during the past three decades. The excursion of its channel has taken place mostly towards east and is in sympathy with the slope of the northern portion. Recently, translation in its course has been most conspicuous as this river formerly flowed in the neighbourhood of Madhubani town in 1950 but after 1960 floods the river all on a sudden took a rapid swing approximately 19 km. eastward to join the Soni stream near Pipraghat before emptying into another powerful stream, the Balan river. The combined course is now known as the Kamla-Balan.[6] The Kosi's westward shift, on the other hand, has been rather spectacular in the sense that in course of its westward translation, it usurped the Tiljuga and finally entered into the newly created Madhubani district.[7] Now almost the entire fringe area of the region is covered by the Kosi flood plain where its channel is more or less stable after the construction of flood embankment since 1955. The Kosi forms an arc of a circle and flows through a circuitous route along the Madhubani-Saharsa district border and then flowing further south it again turns to southeast and east until it meets the Ghugri Kosi near the Mansi junction. From here onward it runs almost parallel to the Ganga in a well defined channel as far east as its confluence with the latter. In the south-central part too rivers like the Baghmati, Kareh, Burhi Gandak, etc., flow through winding courses until they make their way further southeast to the Ganga. The depositional pattern consists of the radiating distributary of streams of which now one, now another will dominate in turn so long as deposition still causes a change of direction as in the Kosi. Associated with this are the spill banks, spill slopes, and spil hollows created in the wake of annual flooding and westerly translation.[8] According to Buchanan, the geographical nomenclature of the streams is to the last degree confused for the river is known by the same name, e.g., Kamla, Tiljuga, etc., Over distances situated miles apart and the same river may be called by different names in different localities of its longitudinal course or two rivers flowing as independent channels may be called by the same name.[9]

TABLE 5.1: **Approximate Rate of Shifting of the Kosi**

Year	Years Interval	Distance moved in km.	Per cent of Movement
1736-1770	34	10.72	10
1770-1823	53	9.28	8
1823-1856	33	6.08	5
1856-1883	27	12.80	11
1883-1907	24	18.40	16
1907-1922	15	10.88	10
1922-1933	11	28.80	25
1933-1950	17	17.60	15
1736-1950	214	114.56	100

Source: Data based on Research and Investigation Division: River Valley Projects Department, Kosi Wing, Birpur (Saharsa).

Further, as is to be expected the snow fed Himalayan rivers coupled with their tributaries and distributaries forming the entire Ganga Basin is influenced by seasonal rhythm, more particularly by plentiful monsoon rains that occur during June-October. As a result, flooding over a wide tract of this level plain and formation of *Chaurs* are most obvious consequences. Floods of very serious proportions have become almost an annual feature in the region. A relatively small area in the plain has to drain off waters of a huge Himalayan catchment in the north with a steep gradient. The catchment basins of most of the Himalayan affluents, e.g., Kosi fall in the area of heavy precipitation. Aided by a fairly good amount of snow-melt water, scouring action in the less stable beds of rivers are found choked with boulders and stones in the upstream courses. These cause impeded flow of water to the Ganga which itself has a large volume of discharge. As the level of the Ganga rises, its northern tributaries do not effectively contain the flow and naturally huge water bodies rushing onto the level plain are unable to be confined to the shallow beds. Thereby calamitous floods, particularly in the southern parts of the plain occur almost every year. Among other factors which contribute to severe flooding in the plain are waterlogging, comparatively high water table in the interfluvial tracts and concurrent translation of shallow river beds, etc.

From an analysis of the foregoing facts of fluvial morphology of the plain it now becomes abundantly clear that most of the rivers descending down the Himalayan slopes are characterised by frequent floods. A Study of the accompanying map (Fig. 5.1) depicting 'Extent and Magnitude of Floods' makes it convenient to divide the whole region into three broad areas, viz., (i) Flood prone, (ii) Flood affected, and (iii) Flood immune. The flood affected zone is mostly found along the left bank of the Ganga and is found throughout along the Gandak, Burhi Gandak, Baghmati, Kamla-Balan courses as far east as the Kosi flood plain. Its width is generally 10 to 15 km. along the major streams. Two of the worst flood-affected areas lie firstly in the neighbourhood of the confluence of the Ganga with the Ghaghra in the west; and secondly, an elongated basin of the Burhi Gandak, Baghmati-Kareh, Baya, etc., which form a zone of convergence of streams in an area of an ill-defined relief, and further east its extension with the Ganga-Kosi confluence. Despite all the flood protective efforts made so far the engineering structures do not have their full impact when rivers are in fury and they act more as a temporary relief to the people and area rather than a permanent solution.

TOPOGRAPHY

The plain on the whole is monotonous and featureless due to low and imperceptible relief. The large scale alluviation and diluviation, meandering of rivers, frequent scouring and erosion of banks and excessive floods resulting into water logging, development of marshy lands (*Chaurs)*, formation of alluvial fans and those of cones and intercones are some of the morphological features that dominate. With the exception of the Himalayan foothills (Siwalik) exposed as bulge in Sumeshwar

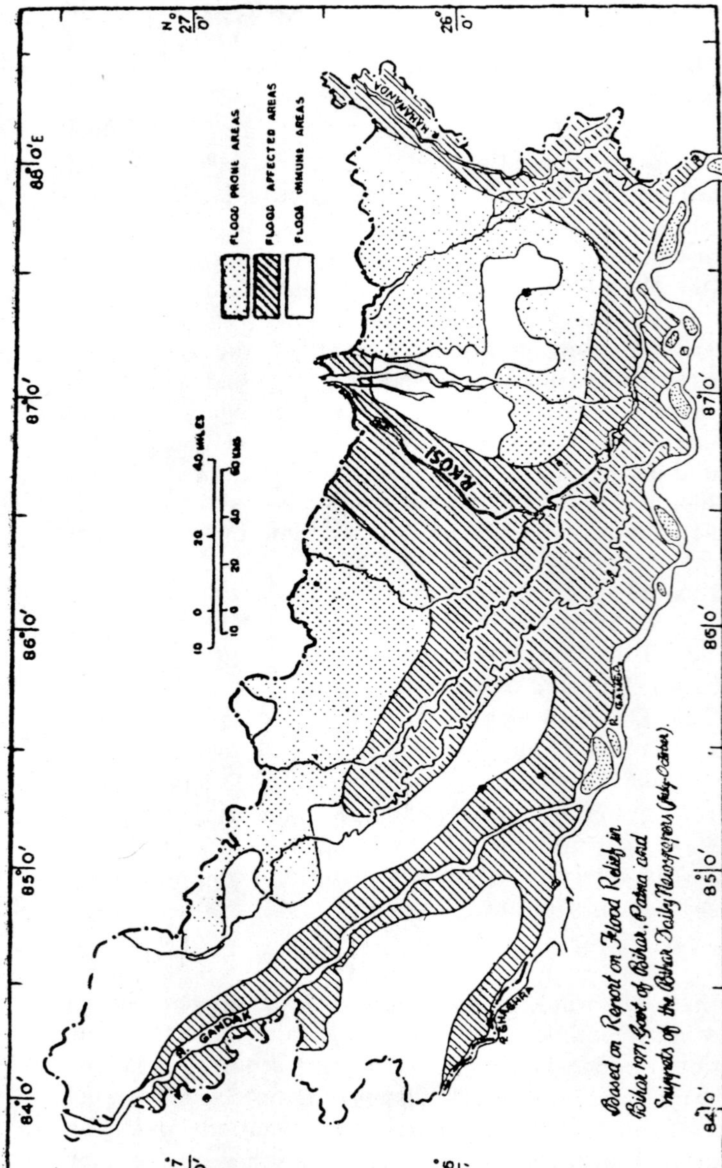

FIG. 5.1 : *North Bihar Plain—Extent and Magnitude of Floods*

and Dun ranges in W. Champaran, almost the entire North Bihar Plain is an area of great physiographic homogeneity throughout its length and breadth from *terai* region of Nepal in the north to the course of the Ganga in the south. Ahmad aptly remarks . . . "the Ganga Plain of Bihar is the superficial expression of a thick body of alluvium which has been gradually accumulating in the vast depression limited in the south by the Chota Nagpur plateau and its outliers."[10] This is a remarkably low and featureless plain composed chiefly of river alluvia laid down by perennial Himalayan streams debouching onto the North Bihar Plain about which we have already discussed earlier. This has often been described as a 'dead level plain' in which it is difficult to detect slope of the area with bare human eyes. However, the level rises imperceptively towards northwest and north where it merges gradually with the foothills of the Himalayas.

(Fig. 5.2) showing 'Drainage and Generalized Contours' indicates that the relief varies from 30 metres in the south-east to 153 metres in the northwest. The general lie of the country is not only from north to south but from northwest to southeast as well. In the central portion the slope tends to be from northwest to southeast. The flow of rivers and streams accordingly bears testimony to these features. Ruggedness number* is the highest (0.24) in extreme northwest (W. Champaran) in the entire plain. In rest of the region it is everywhere insignificant (below 0.06) due to monotonously level land. Consequently, the gradient of the slope is less than 15 cm per km. The drainage courses evidently have a profound influence on topography of the region. In the central north, old courses of the Kamla-Balan coupled with the Adhwara group of rivers have raised the ground level while the eastern portion is a depression in which some of the old alluvial surface has been removed and redistributed downstream. The

* Ruggedness number has been compiled by the authors from Quarter-inch toposheets of the area with the help of the following formula:

$R_n = D_d \times \frac{R_R}{5280}$ Where R_n is the ruggedness number,

D_d is the drainage density, and

R_R is the relative relief.

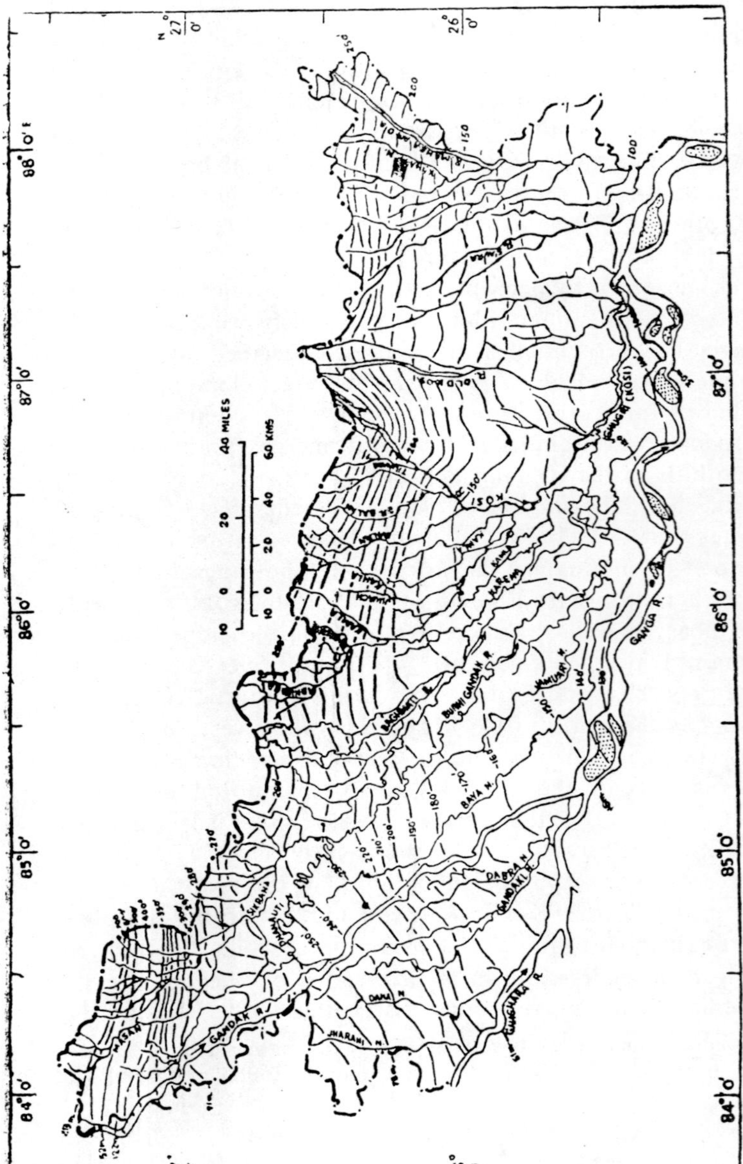

FIG. 5.2 : *North Bihar Plain—Drainage and Generalized Contours*

central linear part is upland where the Gandak, Burhi Gandak, Baghmati and the Kareh run through shifting courses for a considerable distances before their confluence with the Ganga in the east. Similarly, the southwest portion is again an aggradational plain formed by rivers like the Gandak in the north, Ghaghra in the west and Ganga in the south. Apart from these, there lie hilly areas in extreme northwest while the Ganga river imparts its indelible mark on the topography in the southern section.

TOPOGRAPHIC PROMINENCES

The local relief with imperceptible slope differences is profound in this featureless plain. Among the topographic prominences may be listed *levees* (riverside uplands), alternating *Chaurs* (long semi-circular depressions), small ox-bow or elongated lakes, *tals*, dead and defunct courses of ever shifting rivers, meander bends, marshes, bluffs, *diaras*, flood plains and alluvial cones.[11] These clearly show that both the action of alluviation and diluviation are discernible over the surface. All these topographic features merit separate treatment as outlined below.

Alluvial Cones

The cone is a tract of lowland formed by apex in the north terai section where there is sudden break in gradient with a semi-circular fan spreading out to the south co-terminous with the Ganga. This occurs in the fashion of an inland delta. Geddes in an illuminating study has recognised the conspicuous presence of the Gandak cone on the west and the Kosi cone on the east and in between them lie a number of inter-cones that are manifest in the Burhi Gandak and Baghmati river basins apart from the composite Mahananda-Tista cone in the extreme east.[12]

The weathering processes on higher lands are strengthened by the action of gullies and rills, particularly along the *terai* lands of North Bihar. This erosion with the onset of rainy season that corresponds with the *Adra* asterism (last week of June) produces alluvial cones at the base below the boulder strewn debris that is rolled down under gravity in the fashion of an apron[13], especially in the Siwalik hill region of the Champaran district.

The Kosi in course of its long journey through precipitous walls of the Himalayas where it carves out magnificent gorges pursues southward, and after cutting through the Mahabharat Range (Siwalik) between Triveni and Chatra, it emerges onto the Bihar plain in Nepal *terai* and subsequently in Bihar. The bed slope suddenly changes; below Chatra to the south the river currents divide into a number of anastomosing channels, each separated by shoals and bars, and therefore, it is obliged to deposit debris, sand and silt over a wide cone. With apex at Chatra and sides running as far east as Manihari in the east and Khagaria in the west, this river displays more or less, the features of an inland delta. In order that it could adjust itself to slowly changing phenomena of quasi-equilibrium,[14] the river has to pass through the processes of alluviation and diluviation over a great width of its flood plain. By making constant shifts in its bed from east to west, it has formed an alluvial cone. The North Bihar Plain thus consists of cones of the Gandak, the Kosi with two intervening slopes or intercones, lying between them (*cf.* map). The cones exhibit a radial pattern of distributaries, and the intercones generally have lower gradients than the cones.

Chaurs (long semi-circular depression)

The whole face of the plain is dotted with numerous *Chaurs*. It is pertinent to note that at the time of high floods, deposition of silt and sand is accomplished over wide areas. Between the active channels *Chaurs* are so numerous that they break the monotony of the flat plain. The most plausible factor for occurrence of numerous *Chaurs* in the interfluvial zone may be attributed to the fact that in these lowlying areas bounded by natural levees on either side waters rush up and fill up during peak storms and appear here and there in a chain. However, when the flood waters recede, the chain is delinked and the *Chaurs* remain full with sluggish water. The existence of *Chaurs* is particularly conspicuous in the Gandak-Baghmati *doab* as also in the Kosi flood plain to the east. Hardia *Chaur* in Saran district extends for about 32 km. along the Gandak embankment and has a breadth of 3 to 8 km. Similarly there are numerous elongated *Chaurs* southeast off Barauni-Begusarai area. Large and small *Chaurs* are often interspersed with water needs, tamarix and wild grasses along river beds and they are a storehouse of fresh water fishes.

Meanders, Ox-bow Lakes and Bayous

Most of the rivers of the region, before meeting the Ganga, tend to flow in broad sweeping courses. This is due to low velocity of the currents consequent upon deposition alluvium over the vast flat plain that are responsible in diverting the stream courses.[15] Slope and volume of water are other factors which decrease the velocity of current thereby accentuating meandering action. Swing of currents around one curve tends to cross the channel and develops a curve in the opposite bank. Thus meanders grow and shift downstream. The development of length in streams occurs due to meandering which would be apparent from the table given below:

TABLE 5.2 : **Development of Length in Streams due to Meandering**

Between Places	Straight length of rivers in km	Length of rivers with meander km.	Name of the river
Cheria Bariarpur to Khagaria	38.00	88.40	Burhi Gandak
Surajgarha to Bhagalpur	74.80	106.00	Ganga
Chapki to Chautham	72.50	134.20	Kareh

Source: Calculated from quarter-inch Toposheet No. 72/K, Survey of India, Dehra Dun, 1950.

In consonance with the flat gradient of the plain the rivers that flow southwards tend to take pronounced meandering courses. Straleo describes such rivers as so exceedingly winding that lengthening of its course causes meandering.[16] The presence of hard and consolidated blocks of Siwalik clays as buried topography on the banks of most of the rivers has led to the development of meandering courses in this riverine tract. Due to flat and low level of the plain, meandering of river courses has become the principal geomorphic feature, particularly in the lower Burhi Gandak, Baghmati-Kareh and the Ganga. The Kareh—an eastern continuation of the Baghmati makes three meander bends within a short distance of about 3 km. in the district of Darbhanga. Similarly, the serpentine course of the lower Burhi Gandak has given rise to excessive floods during the rains.

Development of meanders leads to the formation of ox-bow lakes and 'bayous' (a name given to the marshy offshoots of lakes). With the growth of meanders, it frequently happens that a narrow neck of land is cut through from two sides. It thus allows the stream to straighten its course and to discharge its flow. There are numerous ox-bow lakes on the bank of the Burhi-Gandak in Muzaffarpur district in addition to a few in the Begusarai-Khagaria area where the Kareh flows down to meet the Ganga southeast off Rosera. When the two ends of the meander bend are separated from the channel, and are likely to be filled up by sand, bayous are generally formed. They lie between the active channels as dead, defunct and mara *dhars*. These topographic forms break the monotony of the flat plain.

Marshes, Diaras and Tals

The abandoned channels of streams in a flood plain may become partially filled with waterborne ready plants, thus forming a marsh. They also occur in a depression on the flood-plain especially in areas of waterlogging. Marshes are numerous towards the south-eastern portion in the neighbourhood of Khagaria-Mansi. Some of the marshy lands dry up during the summer while others exist throughout the year like tanks. Significant marshes are found in the beds of rivers like the Kosi, the Mahananda, the Bhutahi-Balan, the Kareh, the Burhi Gandak and the Gandak. Significantly enough, most of marshy lands merge into river beds during the rainy season. Apart from these, a narrow moist *terai* belt in West Champaran at the base of foothills and a sub-*terai* belt of marshy land in the north with intervening tracts of uplands along river banks are noticeable features. They are formed by emergence of water that is soaked under ground as is obvious in the upper foothills gravel zone of Nepal along the Bihar-Nepal border.

The deposits of river-borne materials in the riparian tract are almost an annual feature. Some backwaters of the river are set up annually in the currents. They become sufficiently stationary to deposit certain portion of the sand and silt hold in suspension. The level of the land which is so far nothing but a heap of river alluvia then gradually rises up as the water lying in stagnant pool spreads out over a wide tract of land. The land thus formed is designated as *diara*. The flat *diaras* predominantly of sand and silt are found on the bank of every perennial river in the region and constitute common topographic prominences of the riverine tracts. On the northern side of the Ganga sandy flats are so extensive that it is difficult to cross them on foot. The alluvial deposits lead to a number of topographic features. Raghopur *diara* (165 km.) and Saintia *diara* (7 km.) provide excellent examples of *diara* topography in the region. On the contrary, the *doab* between the Burhi Gandak and the Baghmati in the southeast is an area of low-lying land and is subjected to frequent inundation. The *tals* may be defined as an extensive lowland appearing as a vast sea of placid waters during the rains and exposed as a fertile tract of land pre-eminently suited to *rabi* culture. These are important topographic features along the course of the Ganga. Baraila *tal* provides most conspicuous example.

Natural Levees and Bluffs

The levee is distinctly an aggradational feature where the river works mainly as a depositional agency. During floods relatively large discharge of silt-strewn water as compared to the graded upper reaches fails to be contained in the relatively shallow river bed and consequently spills over the floodplain. Therefore, the maximum deposition from the sediment load of flood waters is found on the bank of rivers which are gradually raised as natural levees or spill-banks above the interfluvial lowland. The width of the levee varies around a kilometre and may vary from two to ten metres above the interfluve. It thus refers to ribbons of the most recent and present day deposition in the aggrading sections of the courses. Due to the formation of levees and infilling along the sand bank of water course with sediments, most of the rivers in the region have raised their beds and flow in normal conditions well above their floodplain. During the rainy season almost all the rivers overflow their levees and cause damaging floods of serious proportions.

River bluffs with steep banks are met with at some stretches of the river courses downstream. These generally occur on the northern bank of the Ganga, which are, to some extent, responsible for easterly turn of the rivers. Thus prompted by bluffs rivers flow parallel to the Ganga for several kilometres in their last journey before emptying into the Ganga river.

Floodplains

The floodplains or flood terraces are generally riparian tracts exclusively alluviated within the channel and adjacent to them lies the land inundated by floods. Floodplains are (i) always of low relief in comparison to rugged areas of northern Champaran, and (ii) are very unequal in their extent and distribution. The widest floodplain is the triangular inland delta of the Kosi and its tributaries in the east and is the worst flood affected area of the region. It is the young floodplain and is flat to generally sloping in which water-logged areas are found at many places. It is the lowland plain while the Adhwara floodplain, lying north of the Baghmati and west of the Kosi, is an upland plain. It has been mainly formed by alluvial deposits of river Kamla and its numerous abandoned channels in the west together with those of a host of the Adhwara group of rivers. The young floodplains here are situated in between the old floodplains of active streams. However, the boundary between the two is always fluctuating. The floodplain during unusual spill-floods is more extensive than the usual riparian tract formed of new alluvium.

Other Minor Topographic Forms

Among minor topographic features may be included spill-slope, spill-hollows, spill-cone, water logging, *jhils* (pronounced depression), ponds, mango and *litchi* orchards, etc. Most of the *jhils* in the region remain flooded usually from July to October and a few of them remain completely dry during three months, i.e., from April to June. Generally speaking, water-logging refers to accumulation and stagnation of water on the surface or in the immediate surface causing injury to plants, building and other structures. The region, particularly the eastern section, abounds in ponds for irrigation and drinking purposes as well as for the cultivation of *makhana* (waterberry seeds edible). Similarly, areas in east, central south and a number of fragmented patches in south-west of the region are noted for mango, *litchi* and banana orchards respectively.

APPLIED ALLUVIAL MORPHOLOGY

We have discussed at length topographic features of this vast plain. These features are of paramount importance so far as their application to human settlements and economic activities over the entire plain are concerned. Alluvial morphology of the plain has fourfold relations, viz., facilitative, permissive, restrictive and prohibitive to land resources and human activities. It determines to a large extent, the land use and cropping pattern, transport and communication and influences the very way of life of the people. The large scale alluviation and diluviation, meandering of rivers, excessive floods, etc., result into destruction of cultivable lands, developments of marshy lands and at times high proportion of current fallows.

The fluvial action of the major streams has shaped the socio-economic fabric of the people in this region. Two facts of paramount importance which have been duly influenced by fluvial geomorphology of this region and which emerge out in bold relief are: firstly, land resources that are the physical heritage of the past several thousand years; and secondly, the pattern of human settlements that are the direct outcome of the agricultural-economic environment. The pristine richness of soils often reviewed with fresh riverborne silt in course of annual flooding has imparted to the landscape a natural resource—almost unparalleled elsewhere—an asset which has acted as a pull factor for dense human settlements prompting fairly intensive land use. The resource endowment of the region, particularly the one related to land use has played a vital role in conditioning the agricultural land use as well as a diversified cropping pattern.

The area is a vast depositional plain of alluvia which abounds in various forms of texture and the most pertinent fact remains as to the process of alluviation that takes place through a host of streams and their tributaries. Due to fluvial morphology and behaviour of drainage courses soil texture is ever changing in nature as the loam changes gradually into silt loam, silt clay loam and even clay loam.[17] The truth of this fact is also corroborated by the report of the Royal Commission on Agriculture in India as in any particular area the variations are usually associated with the depth of soil and lateral changes are not rapid but when subjected to the action of spill-flooding soil texture may vary within the confines of the land of one village.[18]

Alluvial morphology of the region also influences the field pattern, possibilities of irrigation, the size and distribution of holdings, continuity of cultivation and above all, the most important aspect of agriculture, i.e., the crops—their nature, yield and outturn. The plain topography of the region ensures continuity of cultivation, assures extensive irrigation facilities because of low gradient and huge surface as well as under ground water resources. It further facilitates consolidation of the present small and scattered land holdings as a result of which working on forms becomes easier here than what would obtain in plateau and mountainous relief.

The alluvial deposits of North Bihar Plain by virtue of their loose and unconsolidated nature form the best reservoir of ground water. They mainly consist of varying proportions of sand, silt, clay, etc., whereby almost all lithological types are porous which turn out to constitute the most dependable acquifer zone. Geohydrologically speaking, the whole area is rich in ground water resources and enjoys facilities for 'cavity boring'. A considerable portion in northern part of the plain is not yet under the command of any irrigation scheme and the study area is still in its infancy in the development of irrigation though ideal conditions exist for ground water resources as well as for lift irrigation. In the present state of our knowledge the area is poorly served with irrigation facilities and there are reasons to believe that future planning can be done scientifically for the benefit of agriculture and the peasantry. The alluvial morphology of the plain needs to be considered as a promising line of approach to regional planning of this backward region. The oft-quoted line 'poverty in the land of plenty' requires a fresh look at the problems of the area because though endowed with rich resources its people are still depressed.

The influence of alluvial morphology is also clearly discernible in the difference between the cropping pattern of the riverside levee uplands on the one hand the interfluvial lowlands behind the levees, on the other. The levees are marked by such crops as maize, sugarcane, bhadai rice, oilseeds, etc., which prefer well drained conditions. The entire *diara* tract is pre-eminently suited to growing bumper *bhadai* and *rabi* crops which with fairly high yields have made the farmers economically more strong than in any other part of the region. The carrying capacity of the land is exceptionally high as soils are fertile and are annually renewed with fresh silt. The cropping pattern is much diversified here than anywhere else in the entire area. The density of population is high to very high (635-893 persons per km^2.). *Diaras* close to urban centres grow vegetables in abundance. However, if there is an early incidence of monsoon rain the important *bhadai* harvest of maize is lost and even the bumper *rabi* crops, cannot compensate for the loss sustained by it. Further, the southern portions of the Kosi flood plain which fall in the confluence of the Ganga and Kosi, are subjected to menacing floods every year. Since the southeastern area is exposed to flood hazards almost every year agriculture here becomes at times insecure. The yields are low and naturally the density of population ranges generally from low to very low (435-292 persons per km^2). One significant point to mention is that *bhadai* harvest is comparatively less widespread in northern portions as compared to their southern counterparts. The latter area is dominated by sandy loam and other light-textured soils which are eminently suited to maize crop.

Another important point regarding the morphology of the region is floods vis-a-vis agriculture as the vast load of flood water destroys standing crops and kill them. Deprived of air and sunlight on the one hand and labouring under the heavy load of water on the other crops generally wither away. Only some aquatic plants like rice over the entire area and jute in the east bear this strain and suffocation of floods and that too only for a limited period. The lowland or *Chaurs* grow therefore transplanted paddy and such *rabi* crops which derive benefit from the residual moisture of paddy lowlands. By and large, though floods damage human habitations affect agriculture and the livestock in addition to transportation facilities yet floods are a blessing in disguise.

Morphological features profusely influence the pattern, characteristics and problems of communication and transport. North Bihar Plain has a network of roadways and railways on account of plain character of its land. Roads and rail lines facilitate trade besides catering to the needs of passenger traffic. However, significant they may be rivers and streams in the region present difficulties during the rains. Therefore in order to avoid difficulty of constructing bridges and to check interference with drainage, railways often run parallel to the rivers. The number of metalled roads and their mileage is limited due to high cost of construction. Metalled roads run for short distances, joining the important urban and trade centres of the region. The influence of topography here is manifest in the peculiarities of operation and structure of short, narrow and metre gauge railways in the region. Enormous time to cover short distances is another example of adjusting railway engineering to fluvial morphology. The geomorphic control on railways is further seen largely as railless zone in the hilly area in the extreme north west of the region. In regard to airways and waterways the plain is characterised largely by facilitative geomorphology because of flat terrain, low gradient, slow flow of water which ensure large volume of water in the waterways making seasonal navigation, at least possible, etc.[19]

From the foregoing analysis it leaves an impression that morphological characteristics greatly influence the socio-economic and agrarian aspects in the study region. The human settlements, agrarian activities, development schemes and other factors of infra-structure are largely shaped by the topographic expressions in the area.

REFERENCES

1. R.R. Diwakar, (Ed), 1959: *Bihar through the ages*, Calcutta, pp. 24-25.
2. D.N., Wadia, 1966: *Minerals of India*, New Delhi, pp. 9-10.
3. O.H.K. Spate, & A.T.A. Learmonth, 1967: *India and Pakistan*, London, p. 34.

4. L.S.S., O'Malley, 1907: *Champaran District Gazetteer*, Patna, p.10.
5. K.N. Das, 1968: Westward shift in the courses of the Kosi, *NGJI*, Vol. XIV, Part I, p. 28.
6. K.K.L. Das, 1979: *Population and land resources in North Bihar Plain—West of the Kosi*, Bhagalpur Univ. Approved Ph. D. Thesis (Unpublished), Bhagalpur, p. 53.
7. K.N. Das, 1968: Op cit., pp. 26-28.
8. A. Geddes, 1960: The alluvial morphology of the Indo-Gangetic Plain: Its mapping and geographical significance, *IBG Trans & Papers*, Vol. 28, pp. 263-65.
9. F. Buchanan, 1928: *An Account of the District of Purnea 1809-10*, Patna, p. 9.
10. E. Ahmad, 1965: *Bihar—A Physical, economic and regional Geography*, Ranchi, p. 22.
11. K.N. Das, (Ed), 1969: *Essays in geography*, Bhagalpur, p. 91,
12. A. Geddes, 1960: Op cit., pp. 261-62
13. G.D. Despande, and L.S. Bhat, 1934: The Upper Panch Ganga Basin : A study in the geomorphology of the Deccan traps, *IGJ*, Vol. XXIX, Nos. 2 & 3, pp. 3-47.
14. Lund. B., Leopold, Wolman, M., Gordon & Miller, John, P. 1964: *Fluvial processes in geomorphology*, pp.264-74.
15. P.G. Worcester, 1967: *A Text book of geomorphology*, Bombay, p. 200.
16. J.R. Russel, 1954: Alluvial morphology of Anatolian Rivers, *AAAG*, Vol. XLIV, p. 370.
17. T.L. Lyon, et al., 1952: *The nature and properties of soils: A college text book of Edaphology*, pp. 51-53.
18. Report of the Royal Commission on Agriculture in India, Bombay, 1928, p. 70.
19. E. Ahmad, 1979: Geomorphology and communication, Lecture delivered at the *All India Summer Institute on Recent Development in Applied Geography*, Univ. of Rajasthan, Jaipur, pp. 6-8.

SIX

R. K. Rai

GEOMORPHOLOGY AND RURAL SETTLEMENT IN MEGHALAYA

IT IS an established fact that settlement is man's first step towards adopting himself to his surrounding environment. There are several geographical factors which influence the settlements, their location, site and distribution. The geographical factors which are associated with agriculture also have great impact on the nature and distribution of rural settlements. It is observed that rural settlements are directly or indirectly influenced by the nature of terrain, soil, topography, degree of slope availability of drinking water, accessibility to the region and climate of the region. The settlements of even historical past must have been influenced by the nature of terrain, soil and climate.

The present paper is an attempt to establish more precisely a few of the links between the regional geomorphology and location and settlement patterns in Meghalaya.

Location

The Meghalaya State is located in the northeastern part of the Indian territory. It is located (Fig. 6.1) between latitudes 25° 05′ N to 26° 10′ N and longitudes 90° 0′ E to 92° 45′ E. The State is surrounded in three sides by Assam State and in south and southwest by Bangladesh. The total area of the State is 22,489 sq. km. It has been divided into five hill districts. This State is mainly dominated by three hill tribes, i.e. Khasi, Jaintia and Garo tribes.

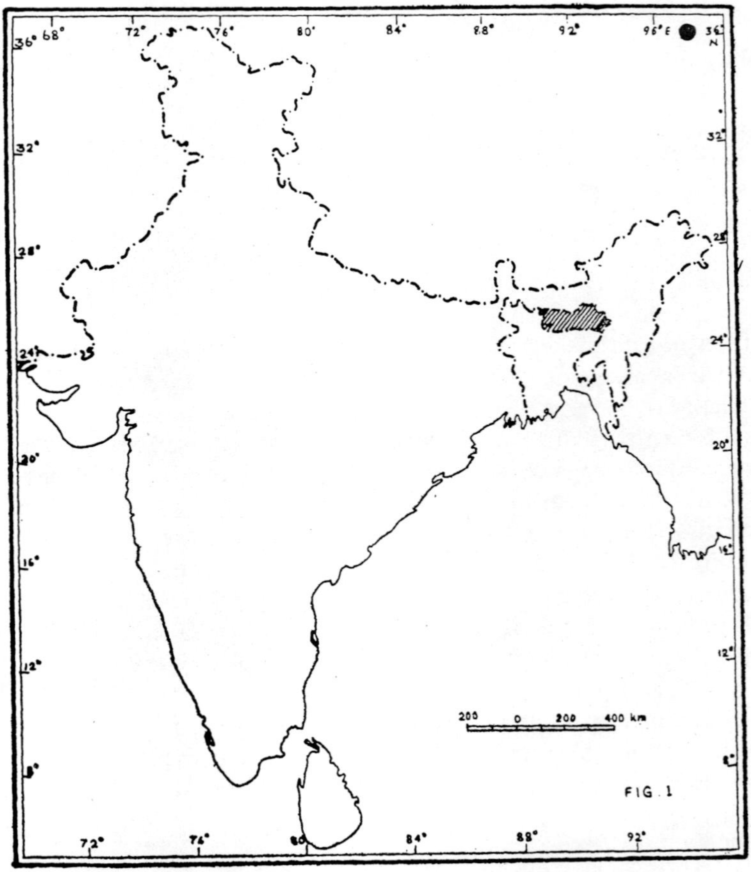

FIG. 6.1 : *Meghalaya Location*

Geology

The Meghalaya plateau is mainly composed of (a) Archaean gneisses with acid and basic intrusives, (b) Shillong group of rocks, (c) Lower Gondwana rocks, (d) Sylhet Traps, and (e) Cretaceous Tertiary sediments.

The distribution of above rock formations show that in central and northern parts of plateau the gneiss and granite rocks are present while in south these rocks are concealed beneath cretaceous and tertiary deposits. In eastern and central parts of the plateau the Shillong group of rocks are exposed.

The lower Gondwana rocks have been recognised in Garo Hills while some deposits of coal have also been traced in Khasi Hills. The Sylhet Traps are exposed along the southern border of Shillong Plateau. While the cretaceous-tertiary sediments occupying the southern part of the Meghalaya Plateau are thick and extensive in nature.

The above geological formations of the region differ in mineral composition and lithological characteristics. The structural characteristics of the region are also quite complex. Number of faults and folds are present in the rocks. The intermittent uplift of the plateau in different geological periods have given rise to a complex multicyclic landscape which are present in remnants of different erosion surfaces in the region.

Geomorphic Features

The Meghalaya Plateau is an eastward extension of the massive block of Indian Peninsular shield. The height of the Plateau varies from 160-1800 metres. The Shillong peak, the highest part of the region is 1957 m above sea level. The middle part of the Plateau in eastwest direction marks the major watershed zone between the tributaries of the Brahmaputra river flowing from south to north and tributaries of Surma valley of Bangladesh flowing from north to south. The height of the Plateau gradually decreases from east to west. The northern hilly terrain gradually slopes down towards the Brahmaputra valley. While the southern part of the Khasi-Jaintia Hills standing in the form of a scarp overlooking the Bangladesh plains.

Due to intermittent uplift of the Plateau and differential erosion the remnants of different erosion surfaces may clearly be traced. There are two major peneplains in the plateau. One extends between Khanpara to Jorhat and the other from Burnihat to Nongpoh. The central upland consists mainly rolling grassy downs intersected with narrow V-shape valleys.

Due to intermittent uplift and active erosion, numerous streams have carved their deep narrow valleys. In most part of the Plateau the process of differential erosion and mass-wasting have either removed the soils entirely or have left a thin residue mixed with sand and gravels. The presence of numerous waterfalls and rapids in neighbouring areas of Shillong and swift flow of streams indicate that this region has a youthful topography which may be due the upliftment of the region in recent past.

The western part of the Garo Hills is a dissected tract with an average elevation up to 600 m or less. The Mones-khola Adaguri range is the demarketing line of western and central part of the Plateau. The northern, western and southern part of the Garo Hills gradually reaches to a plain surface. The bordering areas of the Garo Hill are covered with recent alluvium.

The contour map (Fig. 6.2) of the Meghalaya Plateau indicates that the Khasi and Jaintia Hills are much more dissected into deep narrow valleys and isolated rounded hills. The northern hills of the plateau gradually slope down to-wards the Brahmaputra valley. The southern face of the Khasi and Jaintia Hills stand like a structural platform.

The study of nature of slope profiles of the Plateau clearly brings out the characteristics of slope elements. The nature of profiles clearly indicates that the profiles have gentle to high slope from its crest to foot of the hills. In comparison to other areas in this region three slope elements are clearly visible i.e. the crest, the debris slope and the pediment. The scarp face is missing. The crest is noticed in the form of flat top surface. While the debris slope is the quite extensive part of the hills. This debris slope is covered by the thin soil transported from the crest of the hill. The thickness of the

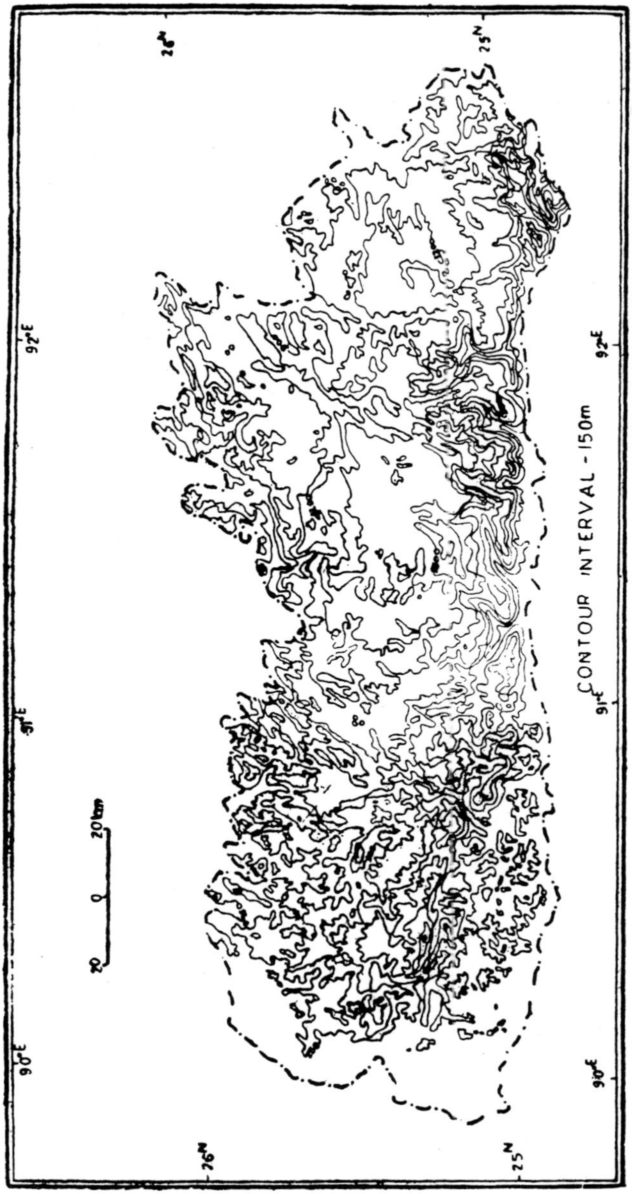

FIG. 6.2 : *Meghalaya Contour Map*

soil gradually decreases towards the crest. The pediment surface is also developed.

The drainage in the Plateau presents a most spectacular feature revealing an extraordinary straight courses of the streams. The magnificent deep gorges scooped out by the rivers in southern Khasi and Jaintia Hills are the result of massive headward erosion by rivers along joints and cracks. The intermittent uplifts of the Plateau have also helped in deepening the valleys. The drainage system of the Plateau is characterised by radial and rectilinear drainage pattern. Thus the topography of the region is of multicyclic nature.

SETTLEMENTS

The location of settlements is an expression of the geographical factors favourable at a particular spot on the landscape. In an agricultural country like India the main factors which determine man's choice of a site for settlement is the nearness to farm, availability of drinking water, a high ground from its surroundings, accessibility to nearby places and climate of the region.

Size and Distribution of Rural Settlements

There are 4,583 inhabited villages in the State with rural population 8,64,529 persons according to 1971 census. The density of population is 45 persons per sq. km. The average size of the village is 189 persons. The average size of Garo Hills district village is 167 persons and for Khasi and Jaintia Hills district is 211 persons. Thus in comparison to plain areas the size of village in Meghalaya is too small.

TABLE 6.1 : **Number of Villages According to their Population size in Meghalaya (1971)**

Population per village	Number of villages
Less than 200	3192
200—499	1093
500—999	237
1000—1999	54
2000 and above	7

Thus there are 69.6 per cent villages with less than 200 population while 23.8 per cent villages have population between 200-499. There are only 5.17 per cent villages with 500-999 population. There are only 61 villages with more than one thousand population in State.

Location and Growth of Villages

In Meghalaya Plateau only 15 per cent land is available for cultivation. Due to dissected terrain and thin soil cover the remaining land is not suitable for settlements. Because of its difficult terrain, the Plateau was thinly inhabited in its historical past. Even today, the only three tribes Khasi, Jaintia and Garo are dominating in the State.

In contrast to plain areas the rural settlements in Meghalaya are confined to plateau tops, gentle slopes of the hills and intermontone valley areas suitable for agriculture. A few households usually controlling large area due to less productivity of land. In Meghalaya still nearly 40 per cent population depends on jhum cultivation. In Garo hills the size of villages is very small. Some of the villages have even less than 10 persons. Most of the population live by shifting cultivation for which only a few of them live in a village keeping around them big jungle area for jhumming. This is also true to other parts of the State but to less degree. Because of shifting cultivation, the villages are not of permanent nature.

The study of one inch toposheets and field study of villages of Meghalaya reveals certain types of settlements. The factors which influence the settlements of this region are the topography, climate and material culture of the region. In the State mainly scattered and dispersed settlements are found. Although along roads linear settlements may also be seen, the scattered settlements are found on hill tops and dispersed settlements are found along hill slopes and river valleys.

The settlements of the Meghalaya Plateau may be classified in the following types :

(1) The settlements on the Plateau tops.
(2) Settlements along hill slopes.
(3) The settlements along river valleys.

The Settlements on the Plateau Tops

The term plateau tops itself indicates that these are the highest parts of the plateau. Although the height of these plateau tops gradually decrease in height from east to west. The field study and one inch toposheets show that the settlements are of scattered nature. A small number of huts are scattered. The density of population is also low. Due to the practice of jhumming the settlements are not of even permanent nature. Thus the cause of scattered nature of settlements may clearly be analysed in terms of the geomorphic characteristics of the region which has great impact directly or indirectly on the settlements.

Much of the topography of plateau is dissected into isolated hills. These hills are separated by deep narrow valleys. Thus these isolated hills have limited land for cultivation. The surface soil is also inferior and less productive. In rainy season much of the surface soil is washed away by sheet and gully erosion. The practice of jhum also exaggerate the soil wash. The accessibility to these areas is also difficult. As these areas are cut off by the presence of deep valleys and covered with thick forest, so the approach from one place to another is quite difficult. The source of drinking water is also an important consideration for settlements. The main source for drinking water is nearby springs or nalas. These settlements are located on raised and prominent sites from where these could command a panoramic view of the country and protective watch over the cultivated fields. Most of the tribal settlements have locally difficult approaches. They are minimally connected with each other by footpaths. Only in recent time a few metalled roads cross the area. Due to jhumming practice the rural settlements were not of permanent nature. Generally the tribals were moving every year from one place to another in search of new land for Jhum cultivation. Due to social customs of the tribals in Meghalaya the most land belong to community and people are free to cultivate wherever they want and as much area as they want. But now due to increase in population there is a pressure on land. So jhumming is gradually decreasing year after year. So now rural settlements are of permanent nature. In Garo Hills also the hill top settlements are of scattered nature.

The important rural settlements in Meghalya located on plateau tops are Mokalaw, Rymbai, Pambrew, Darugiri, Darakgiri, Songmagiri, Nongstoin, Mawphlong, Cherrapunji.

Settlements along Hill Slopes

In State like Meghalaya where the terrain is very much dissected and available land for cultivation is limited, the settlements are forced to develop along hill slopes. The study of one inch toposheets of the Meghalaya clearly shows that next to river valleys the second most important site for settlements is along the hill slopes. The degree of slopes varies from place to place. It is seen that agricultural activities take place at low angles but jhum cultivation is done even up to 70° or more.

The hill slopes are covered by thin layer of soil which gradually decreases as the angle of slope increases. At low angles paddy and maize. etc., are grown but at higher angles inferior types of crops are grown. Along hill slopes terrace forming is an important type of agriculture. The sloping land has been cut into terraces and on these flat terraces paddy crop is grown. The terrace cultivation also helps in checking the soil erosion along slopes. Along slopes the settlements are scattered more or less in linear fashion. The source of drinking water for the settlements are the springs or nearby streams. Along slopes the cultivation depends on natural rains. As these hills are separated with each other by the presence of deep narrow valleys. So there is communication gap in the region. It is difficult to reach from one hill to another hill. So the development of transport network is of great importance for the development of the region. The important settlements along hill slopes are Marngar, Saipung, Pengdim, Khalu, Mibonpara, Jangrapara, Kyrdem.

Settlements along River Valleys

In hilly State like Meghalaya the land locked areas are along river valleys. These are more fertile areas where soil has accumulated from adjoining hills. The accessibility from one valley to another is also rather easy. So the concentration of population is more in comparison to hill slopes and hill tops. The number of settlements along river valleys in Khasi and

Jaintia Hills are more than in Garo Hills. The source of drinking water is nearby streams. In some places the irrigation is done by the streams. The valley areas are more connected by roads. So the settlements are big in size and are of permanent nature. The yield of crops per hectare is also more in comparison to hill slopes and hill tops.

The study of one inch toposheets of Garo Hills clearly shows that the number of settlements along its northern, southern and western borders are more. The border areas of Garo Hills are plain areas with thick alluvium soil. The size of the villages are also big. Along roads linear-cum-scattered settlements have developed. The important settlements along river valleys are Umlong, Lilang, Rangra, Kherapara, Jarain, Burnihat, Umran, etc.

CONCLUSION

The study of geomorphic characteristics of Meghalaya Plateau and rural settlements brings to the conclusion that there is a close correspondence between the geomorphology and rural settlements. The dissected terrain with slopes even more than 70° has forced to adopt the jhumming cultivation. Although this may be due to the socio-cultural traditions of the tribals which dominate in the region. The remnants of younger erosion surfaces are of more concentration of settlements. While the older surfaces have scattered type of settlements. The footpaths or roads have developed along river banks. The dissected topography has posed a serious problem for the development of road and communication lines.

Despite the fact that agriculture is the main occupation of the state but it still faces food deficit mainly because of low proportion of cultivable land and low productivity particularly on hlll slopes and hill tops. The physical conditions of region determine certain special feature of agriculture and economic activities of the people. Gentle slopes upto 30° with good soil depth may be terraced for cultivation. The areas of steep slopes could be brought under forest to check the soil erosion and to increase the income of the state.

SUGGESTED READINGS

Geological Survey of India, 1974: Miscellaneous publication No. 30, *Geology and mineral resources of Meghalaya.*

Rai, R.K., 1976: Geomorphology and rural settlements in the Sonar-Bearma basin (M.P.) *Essay in Applied Geography*, Prof. S.M. Ali Memorial Committee, Geography Department Saugor University, Saugor.

District Census Handbook, 1971: United Khasi and Jaintia Hills, Shillong

District Census Handbook Garo Hills, 1971: (Shillong)

SEVEN

B. L. Sharma

GEOMORPHIC CONTROL ON AGRICULTURAL TAXONOMY: A Study in Rajasthan

INTRODUCTION

WHETHER it is latitudinal location or stage of economic development including technological inputs, the geomorphology influencing the entire agricultural structure. 'By determining the gradient aspect, insolation and radiation, soakage and run-off of precipitation, soil characteristics and wind forces and direction, geomorphology influences the type, growth, output and quality etc. of agricultural crops'. (Enayat Ahmad).

Farmers are no doubt decision-makers and what commodity out of the practicable range will be grown or produced from the field is the result of the past inherent experiences with which they are contended in a set of natural environment. Even then the various agricultural systems and traditional practices of farming in different parts of the region are very largely dominated and controlled by the geomorphic forces and processes. The agricultural systems like rotation of crops, fallow system, intensity of crop land use, open and closed field systems and livestock breeding etc. are conspicuously controlled by the geomorphology of a region.

PART A : GEOMORPHIC SETTING

PHYSIOGRAPHY OF THE REGION

Western Rajasthan

Physiographically, Rajasthan desert, which stretches from the Indo-Pakistan border in the west to the Aravallis in the east and from the plains of the Punjab in the north to Run of Kutch in the south. It is an undulated series of sand ridges, dunes and hillocks. The various factors like topographical and meteorological conditions, resulted in the wide expanse of sandy waste interrupted by rocky potrugens of jurassic, cretaceous and eocene times, process its marine existence over silting of Himalayan rivers. In general, the eastern part of this region is of an undulating nature, while further west it becomes almost a level plains with a few hills. Topographically it ranges between 30 metre to 350 metre above MSL in height. The sand of this great desert is formed out of (i) subaerial denudation of rock, (ii) aeoline source, and (iii) ancient coasts of estuaries.

Aravallis

Extending from Champaner to Delhi in a regular SW-NE direction, the Aravalli is only a stump of a worn out mountain range with numerous discontinuous and echeloned ridges half buried under the desert sand and trans-Yamuna alluvium. The average elevation of this range is more than 600 metres above sea level, acts as a major watershed in two separate drainage systems.

The northern section of Aravalli stretches from the low ridges of Delhi to the isolated hills of Alwar and Jaipur, where it composed of higher relief near Alwar, Khetri, Sikar and Jaipur. The fluvial courses are mainly aligned to synclinal structures and zones of softer phyllites, depicted in a more mature topography. The general elevation of this region goes up to 400-600 m above MSL.

The central part of Aravalli extends from Sambhar Lake to Bhorat plateau south of Deogarh peak. The Northern part is well marked with a saddle in between Jaipur-Alwar hills and

Ajmer-Mewar. The most interesting point is of the existing several wind gaps and passes that provided the passage between east and west Rajasthan. These wind gaps are considered as the remnants of ancient fluvial courses across the Aravalli range, which at a later stage, due to aridity, the extinction of fluvial cycle and sprayed of fine sheet of dust and sand over these areas, checked the courses of ancient drainage and some existing lakes.

The third section of Aravalli range stretches around Udaipur and Mount Abu (1694 m above MSL), structurally and morphologically, this part is much complicated and diversified in nature, composed in high hills and plateau regions. This region is characteristically made up of Aravalli quartzites and lime stones of Raialo series. The western slopes are more steep than eastern Aravallies. On the eastern fringe of the desert is scrapped by typical Aravalli which attains the height of 1500 m above MSL at Mount Abu. Towards Udaipur it stretches to 1050-1200 metres.

Eastern Rajasthan

The area east of the Aravalli in the districts of Jaipur, Alwar, Bharatpur, Ajmer, Sawai Madhopur and Tonk is largely on alluvial plain, while southeastern part of the state is constituted the plateau region comprising the districts of Kota, Jhalawar, Bundi, Chittorgarh and Bhilwara, with varying heights.

GEOLOGICAL STRATIFICATION

During the last two decades, Rajasthan has emerged as a significant state on the geological map of India. This is a fact that fairly thick marine self deposits ranging in age from lower jurassic to middle Eocene, occur extensively in the vast peneplaned, sand covered tracts in Western Rajasthan. (Bhatia, p. 889). The two distinct geographical regions dipicted the clear formation of stratigraphy; one to the west and north-west of the Aravalli range and the other to the east and south-east of it.

'The greater part of the trans-Aravalli region is covered by alluvium and wind blown sand and 7 per cent of the total area is occupied by rock out-crops, confined to the districts of Bikaner, Jodhpur, Jaisalmer and Barmer' (Bhatia, p. 886).

The rock formations in this region is composed of Malani-volcanics, the upper Vindhyan sandstones and limestones with fluviatile and marine jurassic sediments etc. While in the southeastern part the Aravalli precambrian metasediments, the Vindhyans and the infra-trapeans, with Deccan traps are the main occurrences. This fact is very well established by the geologists like Blanford (1877), Oldhum (1886-1893), Heron (1932), Pascoe (1950), Krishnan (1952), Ghosh (1952), etc. etc.. The chronology of this region is also well defined and interpreted by a number of geologists.

DRAINAGE PATTERN

The river system which rise on the Aravallis and on the high lands of southeastern Rajasthan had had profound impact on the surface configuration of the state, stretching back to a very long history of some million years. 'The streams of Western Rajasthan unite to form the Luni river which in times of flood enters the Run of Kutch (Heron, 1938). The catchment area of rivers of this area are of enclosed drainage, and their run off after a heavy downpour, which is unusual and rare, accumulates in local sand dune deposition. In southeast Rajasthan there are numerous rivers which fall into the Banas which itself joints the Chambal and finally find their way to the Jamuna system.

The Banas and Chambal bears certain signs of superimposed drainage following a mesozoic upwarping of the range and the Deccan trap activity respectively (Heron, 1934). These discordant structural relations have no doubt much influenced the neighbouring areas, but the present drainage pattern is almost devoid of any geologic control (Tiwari, p. 995).

CHEMICAL COMPOSITION AND CHARACTERISTICS OF SOILS

Soil Types

The major characteristics types on the basis of their formation and occurrences of soils have been divided into the seven principal groups by Ray Chaudhri (1964). A brief account of important soil groups is provided as below :

Rajasthan, where roughly 3/5th area of the State, 0.2 million sq. km. are under arid climatic conditions, salinity and alkalinity in soils have posed a big problem. Because, accumulation of excess of salt in soils and consequent development of saline and sodic soil conditions are common in arid and semi-arid region of the state. Similar views has been expressed by Mehta (1962) and agricultural department of Govt. of Rajasthan (1963-64 under ICAR Scheme) also gave the account of the distribution of these soils in the districts of Jodhpur, Pali, Ajmer, Bhilwara and Bharatpur. Recently, while demarcating the areas of agro-climatic zones of Rajasthan (Mehta, 1968), the occurrences of saline and sodic soils in different parts of the state, particularly more in the arid part of Western Rajasthan and are less predominant in the higher rainfall regions of Eastern part, are fairly and widely distributed.

Roy, B.B. and Kolarkar, A.S. have thrown sufficient light on the causes of salinity of these soils. They are of the opinion that 'one common cause of salinity in Rajasthan is irrigation with saline under ground water of high salinity and high SAR value. Irrigated agriculture in the arid Western Rajasthan is dependent largely on under ground water resources'. While in the Chambal command area and in districts of Pali, Bhilwara, Kota and other parts of the state, the main causes of salinity development in the soils are the impeded drainage conditions due to physical make-up of soil and low permeability, seepage from the main canals, low lying topographical situation and high evaporation in summer season.

Thus saline-sodic soils are wide spreaded in Rajasthan and causing problems due to irrigation either with saline under ground water or due to rise in ground water table. In the districts of Jodhpur, Pali, Bhilwara, Ajmer and Bharatpur the problem area are mostly saline-sodic with sodium chloride as the predominant salt and high exchangeable sodium content (Rajasthan State Agriculture Department, 1963). Depending on the pH and EC values (electrical conductivity) which are fairly indicative of alkalinity and salinity of soils (Seth, 1967). Rajasthan vary greatly in pH value, with other chemical compositions. In Chambal command area Mehta (1958) studied a saline soil with dominant soluble salt as sulphates of sodium, calcium and magnesium, followed by bicarbonates and chlorides. $CaCO_3$ content in top soil is 2.5 per cent which increases to 24.1 per cent at 70 cm depth; pH value is 8.4 (Kota pH value 8.4 to 9.8 at 60 cm depth) while in Luni basin pH is as high as 9.3 at 60 cm depth.

It is seen that in the case of most of the aeolian sand the CaO content varies from 1.0 to 1.5 per cent, where CaO content decreases downwards, resulting in $CaCO_3$ accumulation or kankar bed formation. But the existence of phosphorus and nitrates have made it fertile for agriculture.

Desert Soil

This soil contain about 90-95 per cent sand and about 5-7 per cent clay with a high percentage of soluble salt and high pH value (Mishra, 1967).

In aeoline sand, the CaO content varies from 1.0 to 1.5 per cent. It is noticed that CaO content decreases downwards resulting in $CaCO_2$ accumulation or kankar bed formation. Nitrates are found low i.e. 0.02 to 0.07 per cent, but together with phosphates it becomes fertile at places.

Grey Brown Soils

These soils are found in Jodhpur, Nagaur, Sikar, Jhunjhunu, Pali, Barmer and Jalore districts. The content of organic matter, nitrogen and available phosphorus in the soils is 0.3 to 0.5 per cent, 0.02 to 0.03 per cent and 0.004 to 0.008 per cent respectively (Jain, J.K., p. 1062). The pH value of these soils ranges between 7.2 and 9.2 and they are calcareous in nature, fertility retains due to nitrates. General dearth and absence of water surface as well as ground (30 metres to 122 metres) has made this region more partch and saline.

Grey and Brown Alluvial Soils

This type occurs in Sri Ganganagar district. This soil in fact is an extension of the soil group of the Indo-Gangetic plain. The most characteristic type of this soil is the nitrogen, phosphorus and organic contents which varies from 0.14 per cent, 0.008 to 0.015 per cent and 0.4 to 0.6 per cent respectively. Thus, agriculturally this alluvial soil is most important for growing variety of crops. The extent of Trans-Yamuna plain is also seen in Alwar, Bharatpur districts with a little difference than those of Sri Ganganagar.

QUALITY OF WATER AND ITS CHEMICAL COMPOSITION

Western Rajasthan

The arid and semi-arid tracts lying on the western side of the Aravalli ranges, faces very low rainfall and only source of surface water is Luni river system. North and west of Luni river, there is particularly no surface drainage in the region. As such the development of ground water to supply water for irrigation and even domestic use is very essential. Most of the ground water is contained in the alluvial formation. The only source of recharge of ground water in the region is by precipitation over the area itself and another artificial source is from the water seeping from the perennial canals viz. Rajasthan canal, Gang canal and Bikaner canal systems.

The various studies made on the basis of effective methods reveals that no surplus water balance exists in the area of Jodhpur, Barmer, Pachpadra, Sri Ganganagar and Bikaner. The records of Jaisalmer and other stations also evidently shows that no surplus water balance is available due to less precipitation in these areas. It is reported that the annual ground water recharge comes about to 22600 Mcft. or 0.52 acres ft. (according to Khosla's method—Mahesh Kumar pp. 597-98). Because of the reliability of the ground water supplies, relatively more attention has been focused on the assessment and utilization of ground water resources in the arid regions of western Rajasthan.

The west and southwest of Jaisalmer and Jalore districts in geological formation are jurassic and form 15 to 40 miles east of Jaisalmer are sand-stone formation and Latur series. They have yielded spectacular amount of water (25000 to 42000 IGPH). In the northwest Rajasthan covering Sikar and Jhunjhunu have got 8000 to 15000 IGPH while eastern Rajasthan comprising of Ajmer, Jaipur, Alwar have the under ground water potential of 25000 IGPH with 45 to 100 m alluvium thickness (Inderpal, pp. 647-48).

Eastern Rajasthan

The well waters of Rajasthan are moderately to very alkaline (Sharma—1963, Darra—1964). The salinity and alkalinity are more pronounced in Western Rajasthan and they go on reducing as rainfall increases on the eastern side. The well waters from Jodhpur and Pali are dangerously high in EC, chloride concentration and SAR (Shankaranarayana et al., 1965). Chemical composition of water in the districts of Jodhpur, Pali, Bhilwara, Ajmer and Bharatpur were studied. The report revealed that the saline, saline-alkaline and alkaline water problem is present in all of them with high pH and EC values, ranging between 7 to 9 and 0.3 to 24.0 Mmhos/cm at 25° C respectively. The sulphates, calcium, chloride, sodium are also present in all these districts (Mathur, Sharma and Ganu, pp. 658-59).

PART B : AGRICULTURAL STRUCTURE

AGRICULTURAL PATTERN IN THE STATE

Agriculture in geomorphic terms can be defined as 'The replacement of the natural vegetation by an artificial plant cover consisting of cultivated plants' (Tricard and Cailleux, p. 206).

The distribution of agriculture cannot be described quantitatively in points, as can mining. The agricultural activities which are usually distributed contiguously over the large extensive areas has been concerned with an analysis of the changing pattern of agricultural enterprise system and activities, through time and has therefore, been involved with a methodology of taxonomic characterisation.

Though, it is difficult to define agricultural regions, but where a single crop dominates a unit, factors much more for a crop region. For that many analytical approaches by large number of distinguished geographers and agricultural scientists including numerous natural scientists have been involved to generalise the cropping pattern and their association with the natural factors like physiography, climate, soils and drainage conditions etc. The geomorphic factors combine together have played very important role in the processes and social systems of agricultural activities in shaping their taxonomic characteristics in one side, and the agricultural enterprise system including cropping pattern associated with the natural factors on the other, which have conditioned the definite enterprise systems in a much more complex way in the state. Thus agriculture is seen with a complex phenomena and function of their closely but widely associated geomorphic control in different parts of the state.

It would be quite in sequence to see the different levels of agricultural set-up in contrast to the physiography, soil and water resources embodying geomorphic control in the state. The agricultural set-up may be summarised in the following heads.

Crop Combinations

As regards the cropping pattern, Rajasthan can be distinctly marked in two major crop regions viz. West—coarse grain region; and East—multi-grain region Western Rajasthan (west of the Aravalli) is primarily dominated by the Bajra, as a single monocrop in majority of the districts, while second important crop is pulses. Sri Ganganagar is an exception to this fact where multi-crops are grown. The Eastern Rajasthan in dominated with wheat, maize, pulses, cotton, jowar etc., etc.

Thus, the state has about 7 types of crop combinations with as many as 57 crop associations in all (Fig. 7.1).

Mono-crop-region is predominantly seen in the western-most districts of Jaisalmer and Barmer, while two crops region of bajra and pulses is seen in Jodhpur, Bikaner, Churu, Jhunjhunu, Sikar and parts of Nagaur and Ajmer districts. Bajra, pulses and wheat representing three crop combinations, is seen in parts of Jodhpur, Sri Ganganagar, Nagaur districts. The four crops combination of bajra, oilseeds, pulses and wheat region is very scatterly distributed in the parts of Alwar, Sawaimadhopur, Jaipur, Jalore, Pali, Kota Dungarpur and Jhalawar districts. (Fig. 7.2).

Similarly, five crops of wheat, oilseeds, pulses, bajra and jowar region is also scatterly seen in parts of Udaipur, Chittorgarh, Sri Ganganagar, Bharatpur, Nagaur, Pali, Sirohi, Bhilwara, Ajmer, Jhalawar and Kota districts. While six crop combinations are found in the Eastern slopes of the central Aravalli region, comprising the parts of Udaipur, Bhilwara, Ajmer, Chittorgarh, Bundi and Tonk districts. The seven crop region found only in two parts of the Jaipur and Udaipur districts.

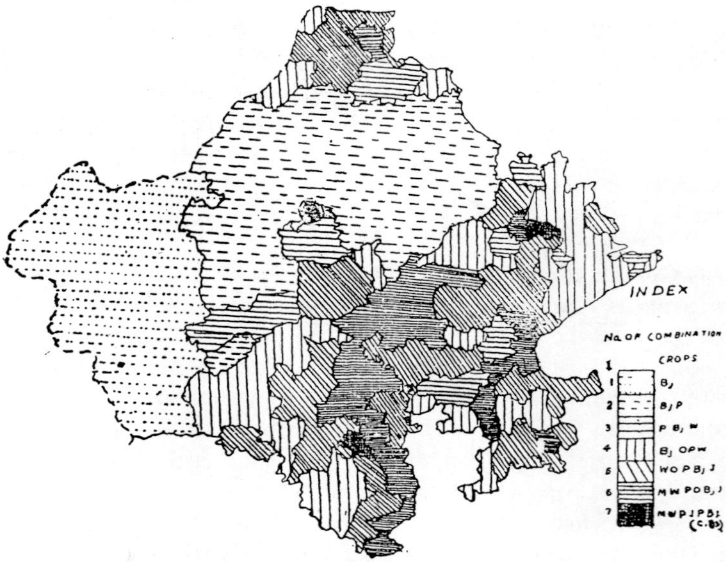

FIG. 7.1: *Crop-Combination Regions*

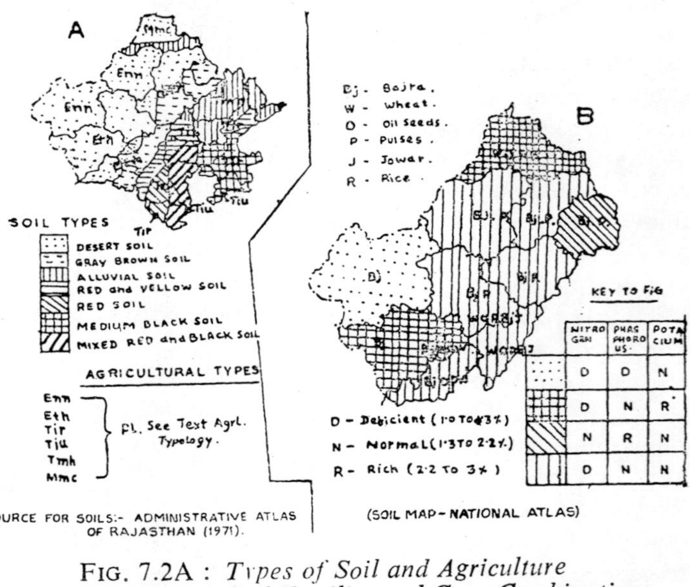

FIG. 7.2A : *Types of Soil and Agriculture*
FIG. 7.2B : *Soil Fertility and Crop Combinations*

Agricultural Typology

To describe agricultural taxonomic characterisation in the state is the study of the agriculture typology based on the I.G.U. Commission's report on the world types of agriculture. Author has tested and analysed the above methodology and has demarcated agricultural typological regions in the state, while completing research on *Agricultural typology of Rajasthan* for his doctoral degree.* These regions, so far studied in the state, are the best classification of their taxonomic characteristic types giving and covering almost all the important attributes of the agriculture. These regions not only characteristic types of agriculture, but show their close association with the geomorphology of the region. Because, they are attributed to relief, soil, water etc. in relation to social, operational, productional and structural principal groups with 27 variables of different aspects of agriculture. On the basis of these attributes, the major agricultural typological regions are determined as (Fig. 7.3).

1. Enn—Nomadic herding.
2. Eth—Current fallow agriculture with migratory herding.
3. Tir —Semi irrigated, low productive, semi-subsistence agriculture.
4. Tiu—Irrigated, medium productive, semi-subsistence to semi-commercial crop agriculture.
5. Tmh--Semi-irrigated, medium productive, subsistence agriculture with livestock grazing.
6. Mmc—Small scale mixed agriculture with crop growing prevalent.

These typological region provide the best account of the taxonomic character in one way, and the control of geomorphic forces on them on the other.

PART C : GEOMORPHIC CONTROL

Geomorphic Control and Extent of Agriculture

With the overall view of the preceding pages reflecting definite formation and direction of geomorphic trend in one side and

* Ph. D. Thesis—'Agricultural Typology of Rajasthan', 1980 submitted to Udaipur University, Udaipur (unpublished).

the agricultural pattern and enterprise systems with types, on the other. The geomorphic control can very well be understood in the State of Rajasthan. The role of various physiographic formations, soil characteristics and water conditions alongwith the other natural conditions like drainage, climate etc. are significant in many respect viz. it has shaped a definite type, character, direction of development etc., etc.

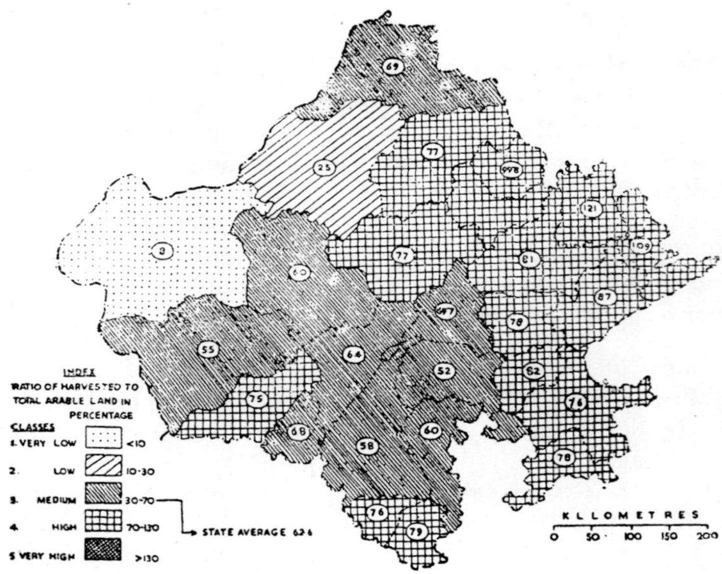

FIG. 7.3 : *Rajasthan State: Intensity of Crop Land Use in terms of the Ratio of the Harvested to the Total Arable Land.*

Physiography has determined their physical limitations and extent of development in one particular direction. In a way it has bounded the limits of agricultural development in various ways and have given rise to a particular system of farming. The technique of agriculture are determined by the development of material civilization and financial resources which in turn are the result of a particular type of economic organisation. But on the contrary geomorphology is so dominant a factor in the human environment that its influence on the agricultural activities is so immense, i.e., in the Aravalli region, the extent and limit of agricultural practices are very much controlled by the relief, elevation slope and geological formation and soil cover. Enayat Ahmad describes this as—"the field pattern refers to the size, layout and alignment of cultivated fields. This is also generally influenced by geomorphology. On level plain surface, fields are rectangular but the rectangulars may roughly be of equal sides. In sloping ground fields are terraced. The terraced walls have to be parallel to the local contour, otherwise the soil and water will not be consumed; they will be washed away" (Enayat Ahmad). Most of the fields found in this region are small in size, terraced in form at the foothills or on the slope. The general nature of the size of holdings in terms of agricultural land, actively employed people and gross output are small. While towards west the size of holdings in agriculture increases to 19.7 ha. in Barmer and 25 ha. in Jaisalmer, no terracing system in general, vast expanse of land but other limitations again control the possibility and feasibility. In the east of Aravalli again the size decreases where farms are found to be 2 to 4 ha. only. In the west people cannot afford to go beyond their control of erratic nature of climate, soil and hazards. While in the central part mountainous region check their extent with slope and thin cover of soil and in the east small patches of field systems.

Physiography has also determined the overall organisations or operational conditions of agriculture in the state. The inputs of animal power and human power is decidedly highly distributed in the west and east respectively. The state as a whole, is dominated by these two power resources in agriculture. Even then, their division is clearly seen where west is dominated by camel as main source of power, with human. While in the east some mechanical power machines can be found standing in the field alongwith animal power to match with the human power. Machines cannot be used in the difficult Aravalli terrain.

Similarly, the intensity of crop land use and the irrigation facilities are very largely indisputably controlled by the physiography in the state. Drainage and irrigation two though different aspects but have the full control of geomorphology as described by Enayat Ahmad "water that flow through rivers may be more easily utilised locally if the level of the stream is not much below the cultivated fields". The state is witnessed very poor irrigation which in turn affect the intensity of crop land use. The western Rajasthan practices long (bush) fallow agriculture, where east with current fallow system, and where water is available either through canals or precipitation, multiple cropping system is found.

Thus, geomorphology influencing the entire agricultural structure makes mountain agriculture different from plateau or plain agriculture. The role of geomorphic control on the agricultural system can very well be ascribed in the words of Tricart and Cailleux (1972). "The cropping system determines the characteristics of the plant cover and consequently, the protection it offers to the soil. This system depends on the economic conditions and the kind of economic organisation. Mono-culture is more apt to produce erosion than diversified cropping, for run off can easily become concentrated and assume increased force on slopes planted in the same crop, whereas with diversifying cropping each field has its own regime, depending on the crop".

In this context state is very largely covered by the mono and two crop combinations in the western sides, where bajra is single predominantly grown over about 2/3rd of the areas with pulses. The eastern Rajasthan (East of Aravalli) is definitely enjoying and well placed in multi-cropping system, hence natural checks and protection to soil erosion is provided. This must have been the reason for depletion of soil cover along the slopes of the Aravalli region.

Soils have the profound impact on the overall agricultural types and enterprise systems in the state. The role of soils in association with the typological regions are the best classification which provide the account of taxonomy and the geomorphic relationship in the state (Fig. 7.2A). Soil productivity of agricultural crops and their range with variety can be ascribed very well in the state (Fig. 7.2B). Soils deficient, normal or rich in nitrogen, phosphorus or potassium directly exert influence on cropping pattern of the area. As has been seen that the larger part of the state are covered by the saline, alkaline and sodic soils which have varying pH and

EC values determine the mono crop or multicropping system at large in the state. Because certain crops sustain to a particular soil characteristic.

The desert soil is characteristically saline in nature, do dot permit the crops to thrive very frequently in the large parts. These are found in Jaisalmer, Bikaner, Jodhpur, Barmer, Churu, Pali, Jalore, Nagaur and parts of Sri Ganganagar districts (Fig. 7.2B). Soils of these areas are characterised as high percentage of soluble salt and high pH value. It has varying percentage of calcium carbonate and is deficient in organic matter, which ranges between 0.2 and 0.3 per cent. This is a part of aeolian sand in which CaO content varies from 1.0 to 1.5 per cent. This region is well known for its mono-culture of bajra and pulses as secondary, grown in rotation to restore soil fertility in some of the districts. Thus the soil plays determinant role in controlling the cropping pattern of this part, which is not at all favourable to grow more crops in absence of water. Poor moisture retentivity coupled with lack of binding material and very meagre bacterial activity render the desert soils highly unproductive. At the intervening long parts of these districts where organic matter, nitrogen and phosphorus in the soil is 0.3 to 0.5 per cent, 0.02 to 0.03 and 0.004 to 0.008 per cent respectively, water is available at the depth of 30 to 122 metres. The fertility of soils is increased particularly towards east and northeast. These soil permits 3 to 4 crops combination of bajra, pulses, wheat and oilseeds etc.

The soils of Indo-Gangetic system as found in Sri Ganganagar, Bharatpur, Alwar and parts of Jaipur, characterised as alluvial of sufficient depth with nitrogen, phosphorus and organic matter varies from 0.03 to 0.04 per cent, 0.008 to 0.015 per cent and 0.4 to 0.6 per cent respectively. This soil is important for agriculture and responsible for the variety of crops such as wheat, rice, cotton, sugarcane and tobacco fruits, vegetable, spices etc. Sri Ganganagar, Bharatpur, Alwar and parts of Jaipur are particularly enjoying better physical conditions in respect of multiple cropping system. These areas are sufficiently growing crops up to four to five combinations in a year. Sri Ganganagar is far superior in this respect where availability of water through Gang canal has placed it at the agriculturally top in the state.

But in certain parts the soils have been worn out by the regular cultivation of particular crops and it affects soils in many ways. "Crop only cover the soil more or less completely; cotton protects it very poorly and corn hardly any better, whereas the small cereals provide a superior cover (Tricart and Cailleux, p. 206). This generalised view of Tricart and Cailleux is quite in relevance to this state also. The coarse grain like bajra, maize and cotton etc. helps in exhaust soil fertility in the west, while in the east wheat, barley, pulses, groundnut, rice, oilseeds, etc., are grown with a sufficient margin of crop rotation where soils are less fertile, and thus protect the soil depletion.

But plateau region of Kota, Bundi and Jhalawar, as well as the Aravalli region is also practising four to six crop combinations where the soils are comparatively fertile, though, calcareous in nature with organic carbon and nitrogen contents. On the Aravalli and on the plateau region runoff is very much checked by the variety of crops grown in these parts in terraces. Runoff can easily become concentrated and assume increased force on slopes if the same mono-crop is planted on it. "Runoff removes soil on a freshly ploughed field, but lower down as the water reaches a wheat field" (Tricart and Cailleux, p. 207).

WATER CONDITION

The cultivation of various crops and the various traditional practices and systems of agriculture are predominantly the result of the geomorphic control of different levels of water in the underground geological formation. The discharge of water at various places in the Rajasthan state reveals that the chemical composition of water and its characteristics feature with varying degree of p^H and EC values, control the cultivation in the area. It has been observed during the wide survey of various parts of the state by Roy and Kolarkar that saline water is due to irrigation with saline under ground water of high salinity and high SAR value, covered by Jodhpur, Pali, Ajmer, Bhilwara, Bharatpur districts. Like soils, the availability of saline water makes cultivation more scanty, poor and precarious as it is seen in the saline and alkaline tracks of western Rajasthan and in parts of Aravalli slopes in south and southwestern Luni basin. Water in the wells sink lower and cultivation becomes more partchy.

In the central parts the pH value ranges between 7.2 and 9.2 and EC 1.14 Mmhos/cm at 40-60 cm, which are giving rise to bajra, pulses, oil seeds and wheat at places. The water requirement for different crops and livestock categories vary greatly in order to survive. It depends on the extent to which structure protects the species from transpiration, since nearly all the additional water required by a growing plant is to replace its essential water content as this is lost by transpiration. Per ton of dry matter produced, cereals transpire 400 to 500 tons of water and grasses more than 800 tons. A field producing 3 tons of dry matter per acre may transpire from 1200 to 2500 tons of water.

Plant vary also in ability to extract water from the soil in favourable condition the root of lettuce penetrates 12 to 15 inches, those of potatoes and peas 2 feet and tobacco 3 feet, maize 4 feet and lucerne and grapes down to 8 to 10 feet or more. For animals, dairy cows naturally have the greatest needs. A Frisian cow yielding 180 lbs of milk per day require about 190 lbs of water. A Jersey cow producing about 30 lbs milk might consume about 100 lbs of water. Beef cattle need about 70 lbs of water. Sheep require about 5-13 lbs (Findings of U.S.A. Department of Agriculture—1955 by Symons, p. 23).

Alongwith plants and various crops which are the result of the limiting soil and water factor, animals also find their feet in accordance with the above stated facts. Large animals like cattle and buffalo are kept in the eastern parts of the state, whereas sheep, goats and camels are the animals of the west. This distinction is made on the basis of the availability of water either in form of rain or surface drainage or even from underground sources. Thus, the livestock breeding system in a way, is also controlled by the geomorphic set-up of the region.

INTERRUPTIONS IN THE GEOMORPHIC EQUILIBRIUM CAUSED BY AGRICULTURE

Alongwith the natural imbalances of resources, the regional imbalances of human activities are seen with varied economy and human needs, caused interruptions in the nature's law. These can be summarised in the following manner.

Of all the man induced interruptions in the geomorphic equilibrium caused by the agricultural activity brings changes to the general landscape. Because of the harvesting and important part of the mineral matter is not return to the soil upon death. Thus, an important part of the mineral element is taken out of the field. This loss is compensated by the use of fertilizers, manures and other improvement but not as sufficient as natural. Secondly, because of clearing the protection of the soil from the elements of the weather is modified. Thirdly, the techniques of agriculture are determined by the development of material civilization and financial resources, depending on the economic organisation. Cost of production and competitive economy give rise to a sort of struggle against nature. For this reason, many new areas have been explored for the cultivation for the future development. Fourthly, the cropping system determines the characteristics of the plant cover and consequently, the protection it offers to the soil. This system very largely depends on the economic conditions and the type of economic organisation. This state is thus, reckoned with the poor economy with traditional subsistence type of farming practising mono-culture or at the most two crops in a year causing much harm to the land. Fifthly, social structure and land system including land tenure determined the division of land and the means available to farmer or peasant to adopt a certain farming techniques. They even influenced the overall economy. Large estates are necessarily a source of revenue, and because of it, tend towards a speculative type of farming. All these systems leading to a serious spurt of erosion and depletion of soil cover. Sixthly, similarly, the use of modern machines in agriculture also causing depletion of soil cover.

CONCLUSION

Geomorphology is so dominating a factor in the human environment that its influence on the pattern and destiny of agriculture is immense. Thus, agriculture in mountains, plateau and plains differ greatly. Similarly the underlying strata of geological formation and surface configuration control the geomorphic processes and their varied impacts are traced upon the human activities. In a way, the saline, sodic and alkaline coarse of soils, underground water of varying pH and EC values, surface drainage with different angles of slopes, not only controls the variety of crops, intensity of agricultural land use, patches of grasses, perennial crops, vegetables, fruits and trees, but also the systems; field pattern, way of life and a definite type of cultivation of crops alongwith livestock enterprise systems. Because, the growth of different type of crops; cereals, fruits, vegetables, spices and other cash crops, vary with geomorphology, particularly slope, plain, plateau and mountain like land features.

SUGGESTED READINGS

Ahmad, Enayat, 1979: Lecture delivered at the All India Summer Institute, University of Rajasthan, Jaipur.

Bhatia, S.B. 1977: Palaeontology of Rajasthan: A review, *The natural resources of Rajasthan*, Vol. 2, Jodhpur University, p. 889.

Blandforc, W.T. 1977: Geological note on the great Indian desert between Sindh and Rajputana, *Rec. Geol. Surv. India*, 10, pp. 10-21.

Darra, B.L. et al., 1964: *Indian Sec. Soil Sci.*, 12, p. 129.

Heron, A.M., 1938: Presidential address, section of geology and geography, *Proc. 25th Indian Sci. Cong.*, pp. 119-132.

Heron, A.M., 1934: *Physiology of Rajputana*, p. 126.

Inderpal, 1977: Water resources for irrigation in Rajasthan, *The natural resources of Rajasthan*, vol. 2, pp. 647-48.

Jain, J.K., 1977: Soils of the Indian desert *The natural resources of Rajasthan*, Vol, 2, Jodhpur University, p. 1062.

Kostrowicki, J., 1976: *World types of agriculture*, Warsaw, Poland.

Mehta, K.M., 1962: Extension and importance of research already done, in progress and required saline-alkali soil problems, *Seminar on salinity-alkali and soil problems*, IARI, New Delhi.

Mehta, J.M., 1958: Report of soil and crop investigation survey, Chambal commanded area (Rajasthan) Department of Agriculture, Rajasthan (Udaipur).

Mahesh Kumar, 1977: Estimation of ground water, *The natural resources of Rajasthan*, Vol. 2, Jodhpur University, p. 597.

Mathur, C.M., Sharma, O.P. and Ganu, S.N., 1977; Saline water of Rajasthan, *The natural resources of Rajasthan*, Vol. 2, Jodhpur University, pp. 658-59.

Mishra, V.C., 1968: *Geography of Rajasthan*, N.B.T., New Delhi.

Nityanand, 1972-73: Crop combination regions of Rajasthan, *G.R.I. Calcutta*, Vol. XXXIV, No 1, pp. 46-60.

Ray Chaudhari, S.P., 1964: The land resources of India, Indian Soils, Their classification, occurrence and properties, *Manager of Publication*, Delhi, Vol.1, pp. 1-4.

Roy, B.B. and Kolarkar, A.S. 1977: Saline and sodic soils of Rajasthan, *The natural resources of Rajasthan*, Vol. 2, Jodhpur University, pp. 1013-14.

Rajasthan State, Department of Agriculture, 1963: Reclamation of saline and alkaline soils in Rajasthan, A reconsolidated report on ICAR Scheme.

Seth, S.P. 1967: Indices for diagnosis of alkalinity and salinity in soils of Rajasthan canal area, *J. Indian Science Soil Sci*, 15.

Sharma, D.L. et al., 1963: *J. of Indian Sec. Soil Sci.*, 11, p. 299.

Shankaranarayana, H.S., et al., 1965: *J. Indian Sec. Soil Sci.*, 13, p. 105.,

Sharma, B.L., 1980: Agricultural typology of Rajasthan (unpublished thesis) Udaipur University, Udaipur.

Sharma, B.L., 1977. Effect of crop combination on concentration and diversification of crop in Rajasthan, *Paper presented in 9th All Rajasthan Geographic Conference at Banswara.*

Symons, L., 1968: *Agricultural Geography*, Bell's, London, p. 23.

Tiwari, A.K., 1977: Geomorphic cycle and land forms, *The natural resources of Rajasthan*, Vol. 2, Jodhpur University, p. 995.

Tricart, J.P. and Cailleux, A., 1972: *Introduction climatic geomorphology*, Longman, p. 206.

EIGHT

Nageshwar Prasad and *Abhimanyu Mahto*

ROLE OF GEOMORPHOLOGY IN REGIONAL PLANNING

WE KNOW that geomorphology tells about the shapes of hills and valleys, the degree and frequency of slopes, the development of drainage patterns and the nature of the material exposed in all available sections (Putnam, 1962). But the role of geomorphology is not limited to this academic significance only. It is true that the cultural value of geomorphology has been underestimated by most people, while the truth is that geomorphology enters into our daily lives in countless ways. The study leads to straight thinking and to more interest in and enjoyment of one of the most important factors in our natural environment. It has, therefore, been rightly remarked that for a better understanding of the environment of which we are a part, we must study the earth, if only to cope with, control, and intelligently manage it (Worcester, 1965; White, 1968; Cooke and Doornkamp, 1974; Jacks and Whyte, 1939; Coates, 1973). Topography obviously plays an important role in determining the most feasible high route, although sometimes cultural features may necessitate routes not in best accord with topography (Thornbury, 1954). Soil maps are, to a considerable degree topographic maps, and the differentiation of the various members of any soil series rests fundamentally upon the different topographic conditions under which each member of the soil series develops. The problem of soil erosion is essentially a problem involving recognition and proper control of such geomorphic processes as sheetwash erosion, gullying, mass-wasting, and stream erosion. Severity of erosion is not determined by the angle of slope alone, but it may be serious on slight slopes where they are on impermeable materials. Similarly sand and gravel have many engineering as well as commercial and industrial uses.

Application of geology to dam-site selection involves a synthesis of knowledge concerning the geomorphology, lithology and geologic structure of terrains. Many of the factors that determine the choice of a site for an airfield are geomorphic or geologic in nature.

Thus, all these conclude that any planning for a region should be conducted in consultation with a geomorphologist so that many problems may be overcome by applying the various geomorphic aspects of a region. A geomorphologist can offer a wider understanding and a greater appreciation of the earth in which he lives.

To elucidate the role of geomorphology on regional planning, here is chosen a problematic region the Barakar Basin of the Chota Nagpur Highlands of Bihar (India) which presents a wide scenery of geomorphic features. For better understanding only one aspect of regional importance is dealt herewith and that is soil and soil erosion.

THE BARAKAR BASIN

Lying between 23° 41′ 30″ and 24° 31′ 30″ N latitudes and 85° 10′ 00″ and 86° 54′ 00″ E longitudes, the Barakar Basin as purely geomorphic unit consists of the whole catchment area (7026.1 km^2.) of the Barakar river in the Bihar State of India. The Basin is a part of the Chota Nagpur Highlands which are the easternmost continuation of the Peninsular upland.

The Barakar Basin is a region of physical inequalities. It presents a rich panorama of topographical features, characterized by scarp-zones, erosional surfaces and residual hills, etc. The general slope of the region is in the southeast direction. . Extensive areas lie between the 152 m and 457 m contours, which present a variety of morphologic features. The 305 m surface constitutes the core of the Barakar Basin. On the southeastern side it gradually slopes down to about 100 m at the confluence of the Barakar with the Damodar river. On the southwestern corner, it ascends up 600 m which is a part of the Upper Hazaribagh Plateau scarps. The transition from 305 m to 381 m surface is gradual on three sides of the plateau.

Thus, the Barakar Basin presents a variety of landscapes characterized by scarps, rugged topography, long, deep and terraced river valleys: waterfalls and rapids, residual hills (monadnocks, tors, etc.), meanders, gullies and other features like elongated ridges and sometimes hilly spurs, projecting from higher table land or elongated into chains of hills and ridges.

The Barakar Basin consists of crystalline gneisses and granites and is supposed to represent an originally solidified crust. The Hercynian revolutions probably resulted in small linear fractures where the Gondwana were deposited. The epeirogenetic uplift during the Tertiary times has been responsible for initiating a new denudational cycle which has carved out the present landscape.

At present the Basin is under sub-tropical humid climate. A large part of the region is covered with rich forest and scrub-jungle. These protect the area at places from severe soil erosion. But great elevations and steep slopes "facilitate quick runoff which renders the streams ephemeral and ponds empty scars in summer" (Ahmad, 1965).

SOILS AND SOIL EROSION

The soils of the Barakar Basin are "residual soils being developed on the underlying parent rocks" (Ahmad, 1965). The rocks which played an important role as parent material in the evolution of the soils of the Basin include mainly sandstone and shale in the Gondwana troughs and granite and gneiss in the remaining part of the Basin. Sandstones give rise to poor sandy soils while shale yields more or less clayey soils which are more retentive of moisture than sandstone soils. Fire clay covers a large area in the southeastern portion of the Basin where it has led to the establishment of many industries associated with it. Sandstone soils are observed in the faulted troughs and river valleys or inter-ridge lowlands. Granite and gneiss have given rise to clayey and sandy and infertile laterite soils. Wherever these laterite soils are "mixed with clayey matter derived from other rocks, they become productive" (Ahmad, 1965).

Thus, in general, the soils of the Barakar Basin are mainly derived from ancient crystalline gneisses: schists and granites which form their bedrocks. These soils are known as 'Red Archaean Soils'. "The high ferric content of the rocks has imparted to the soil a deep red, brown and black colour (Ahmad, 1965). This characteristic nature of soils derived from the rocks of the Basin has direct influence on the agriculture of the country and indirect influence on the human efforts. These soils are comparatively of poor fertility.

We have noted the uplands, lowlands and scarp-zones of the Barakar Basin. On the uplands where there is a great preponderance of thin veneer of red ferruginous, sandy and gravelly soil, fertility depends on closeness to habitation. A large proportion of the ground, both on upland and lower levels, is under terraces given to paddy and here the soil cover is relatively thick, being fertile and more calyey than the eroded soils of unterraced upland. The soils of valleys and depressions receiving all the detritus washed down from the uplands and slopes are relatively moist, more clayey and given to rice. On the irregular slopes of the scarps and immediately below, the soil is composed of infertile debris (Ahmad, 1965).

That the soils of the region are largely a function of geomorphology becomes apparent in the field. On the upland crests there may be no soils and the ground consists only of rocks, hills and hillocks. Lower down adjoining the rocky exposures the soils are thin, sandy, pebbly and frequently strewn with rock fragments of various sizes. As we move downwards the thickness of the soil increases, its texture becomes finer, the colour changes from yellow or brown or red to dark. Sand decreases in proportion and clay and silt increases. Even in the same field the physical and chemical characteristics of the soil depend on geomorphology, i.e., slope and the differential elevation of the field with respect to the adjoining fields.

The need of planning begins with the fact that the Basin has the areas of worst gullies. Gullies or ravines are one of the most spectacular features of the Barakar Basin. One of the peculiarities of gullies is that they are distributed over the whole of the Basin in the undulating surfaces of the west and middle and also in the gentle sloping area of the east, although the degree of their intensity varies to some extent. A detailed discussion in this respect is due to Ahmad (1973), Roy and Bhattacharya (1967) and the report of the D.V.C. (1972). The author has also gone to the field study and has measured the magnitude of ravines at places.

Gullies are found both in the upper and lower reaches. They have been observed to advance by development of side branches. But the encroachment on the uplands is not necessarily preceded by finger-gullying at all the places. The common feature is that vertical clefts develop on the uplands (61 cm to 92 cm) away from the gully head parallel to the perimeter, the soil becomes loose by the rains and the whole column overfalls in the depression. The rate of expansion of the gullies, however, depends mainly on the watershed conditions like slope, vegetation, soil and the length of the slope.

Consequent upon the gullying on the farm lands (uplands) the area under sheet erosion in farm lands has also been proportionately getting reduced. But so far as the areal spread is concerned the reduction is almost negligible in view of the small area of the newly cut gullies. But the expansion of the gullies as such, however, small it might be, is highly significant in view of its destructive nature and the gullies being the major contributor of the silt load of the streams. Sheet erosion is observed to make no fresh headway in the protected areas of the farm lands, i.e., the bench-terraced paddy fields. On the contrary, a small but definite trend is noticeable about converting some of the severely eroded patches into paddy fields by bunding and levelling.

Now when a geomorphic interpretation of these maps (Figs. 8.1/8.2, Table 8.1) is made it becomes clear that the gully erosion is not merely related to the topography, slope, geology, soil, etc. as noted by most of the soil scientists, and also by DVC (1972). We find that the intensity of gullying does not vary markedly from the upper catchments to the lowers, although undulations and slope vary apparently. So we, cannot get inclined to the general views. It is true that topography, lithology, drainage and human interference have important roles, but they are definitely secondary roles. The continued regional uplift or rejuvenation has the primary

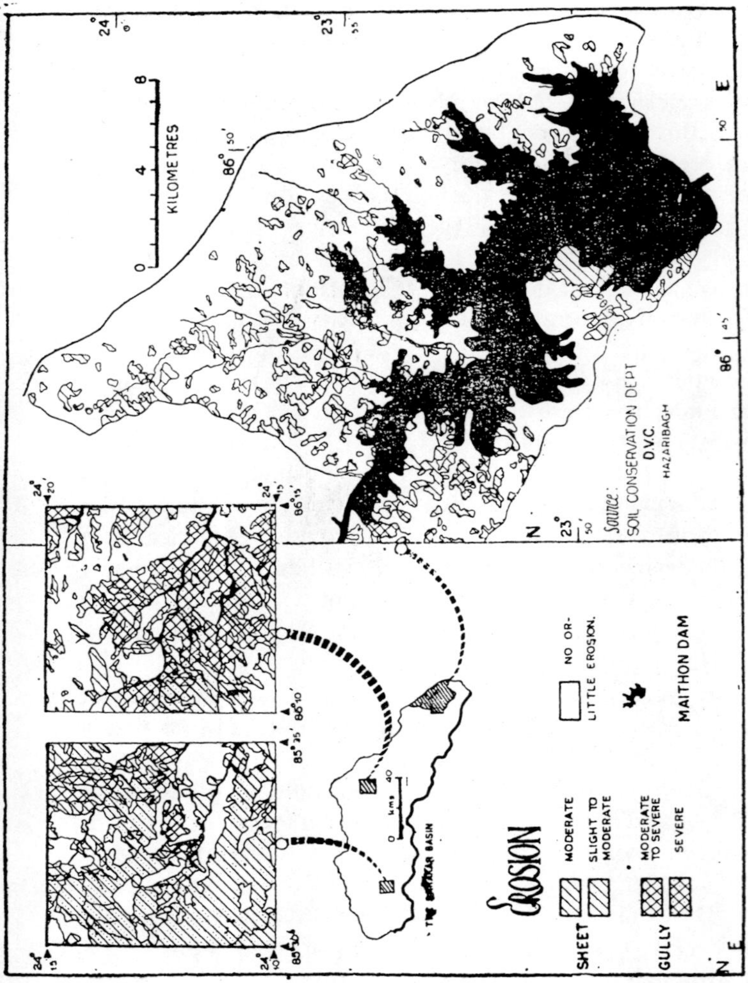

FIG. 8.1 : *Gully and Sheet Erosion*

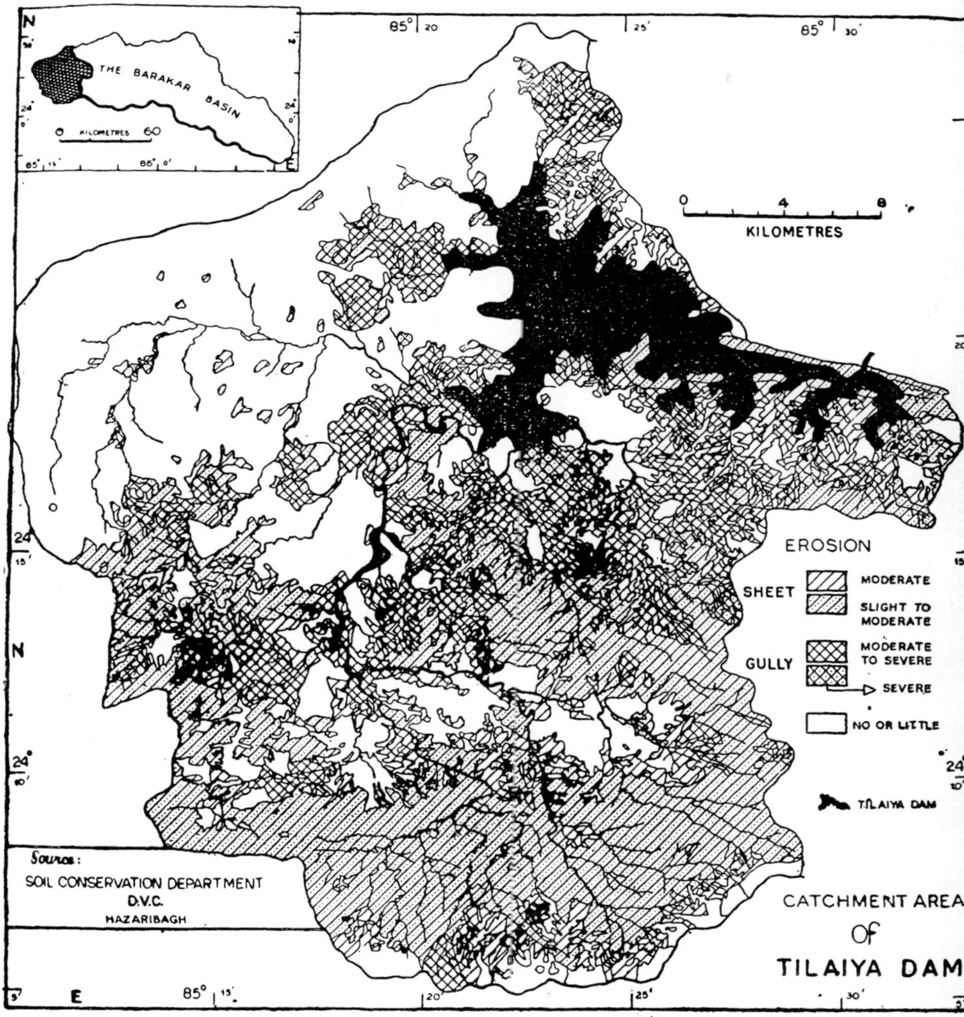

FIG. 8.2 : *Gully and Sheet Erosion*

TABLE 8.1 : **Detailed Statement of Erosion, Dam-wise (The Barakar Basin)**

Sub-Catchment	Land area in acres	Land (in acres) eroded, eroding or subjected to erosion					
		Sheet Erosion in Farm land	Sheet Erosion in Forest land	Total Sheet Erosion	Fully Erosion outside forest	Fully Erosion within forest	Total gully erosion
		TILAIYA DAM					
Saghar-Karkara	44400	3800	18350	22150	6050	1200	7250
Barakar-Kewta	150000	16700	25750	42450	20100	7000	27100
	194400	20500	44100	64600	26150	8200	34350
		MAITHON DAM					
Harnaro	95450	12700	1450	14150	10650	200	10850
Keso	54900	10000	3450	13450	4650	—	4650
Barakar Right Bank	44300	8000	7100	15100	4650	2400	7050
Barsoti	144850	19150	14300	33450	8100	10050	18150
Bareto	93100	16400	5650	22050	4950	50	5000
Irga	97650	23650	10650	34300	19800	1850	21650
Usri	214800	38250	29000	67250	31500	600	32100
Khero	82900	11550	8500	20050	10550	3800	14350
Chirki	156650	17300	43800	61100	4400	3400	7800
Barakar Left Bank	58900	3850	18100	21950	3350	—	3350
Beri	62350	12800	9300	22100	3500	50	3550

Rajoya	73450	15900	3250	19150	3900	—	3900
Barakar Right Bank	77250	8350	9840	18200	2200	1350	3550
Barki Jhar	56000	5150	50	5200	700	50	750
	1312550	203050	164450	367500	112900	23800	136700
		OUTSIDE DAMS					
Khudia Pussi	145550	12250	4300	16550	1400	200	1600

All acreage figures rounded to nearest 50 acres (1967).

role. Thus, we are inclined to the views of Ahmad (1973) who has dealt with a most convincing factor, i.e., tectonic uplift which has been concluded after a detailed study of the gullies, confined mainly to four pockets in India (Sub-Himalayan Punjab, Chambal-Yamuna, area, Gujarat and Chota Nagpur). the conclusion is no doubt, thought-provoking and non-challenging, at least at the present state of study. It will be worthwhile to quote his words, "Even the steep slopes of the Himalayas or the scarp-zones of Peninsular India are free from gully erosion. The peneplaned surfaces of the peninsular tract are also free. No doubt some of these surfaces are densely wooded; others are deforested and bare of vegetation cover. Some are of high steep slope, others are of gentle inclination. These gully-free tracts are varied as regards lithology; structure; altitude; climate; depth; texture and structure of soils; intensity of land use; density of livestock and human population, and intensity of cultural occupance, etc. but they are alike in being free from gully erosion driving us to the conclusion that there is some common cause in the ravined tracts, and this common cause appears to be uplift (Ahmad, 1968).

In environmental management the rate of soil erosion has to be assessed and such an assessment is tied to geomorphology.

The severeness in the gully erosion causes poor return to the people of the Barakar Basin and hence it should be a matter of great concern to the Government. Agricultural planners generally do not know much about the geomorphology of the area, without which they cannot be aware of the important and real related problems. Soil conservation measures may, therefore, not succeed. To check soil erosion, intensive geomorphic survey of the area is essential. Checking soil erosion is a problem largely of adjusting the flow of water and dissolved soil particles to the geomorphology of the land surface. Soil erosion is *inter alia* related to slopes, breaks of slopes and knicks and surface undulations. This makes detailed geomorphic survey and mapping as an essential condition for soil conservation and management.

Ahmad is of the opinion that gully erosion will certainly be checked by (i) fencing off the areas through barbed wire and iron pillars, etc. and allotting them to the landless people who can have love for the land*, (ii) crop rotation, (iii) contour cultivation, (iv) strip cultivation, (v) bunding with grass, (vi) avoidance of fallowing, (vii) ploughing up of incipient ravines and other measures.

SUGGESTED READINGS

Ahmad, E., 1965: *Bihar, A physical, economic and regional geography*, Ranchi University, pp. 54, 57, 60, 66.

Ahmad, E., 1968: Distribution and causes of gully erosion in India, *Selected Papers—Physical Geography*, Vol. 1, 21st Int. Geog. Cong. India, Nat. Committee for Geography, Calcutta, p. 3.

Ahmad, E., 1973: *Soil erosion in India*, Asia Publishing House, Bombay-1, p. 101.

Coates, D.R., 1973: *Environmental geomorphology*, State University of New York. pp. 262.

Cooke, R.U. & Doorikamp J.C., 1974: *Geomorphology in environmental management*, Clarendon Press, Oxford, p. 104.

D.V.C., 1972: *Report on Soil conservation working Plan*, DVC, Hazaribagh, pp. 370.

Jacks, G.V. & Whyte, R.O., 1939: *The rape of the earth—A world survey of soil erosion*, Faber, London; pp. 313.

Putnam, D.F., 1962: Geography is a practical subject, *Geography in the 20th Century*, (Ed. G. Taylor) Methmen, London, p. 399.

Roy, D.K. & Bhattacharya, P.B., 1967: Use of air photographs in assessing erosion in Upper Damodar Valley, *Jour. of Soil and Water Conservation in India*, Hazaribagh, Vol. 15, Nos. 3 & 4, pp. 1-9.

Thornbury, W.D., 1954: *Principles of geomorphology*, John Wiley & Sons, p. 571.

White, J.F., 1968: *Study of the earth*, Prentice Hall of India, New Delhi, p. 2.

Worcester, P.G., 1965: *A text-book of geomorphology*, Affiliated East-West Press, New Delhi, p. 3.

NINE

Donald R. Coates

APPLIED GLACIAL GEOMORPHOLOGY

INTRODUCTION

THE PRACTICAL applications derived from glacial studies differ in at least one significant manner from those of other surface processes, such as, rivers, mass movements, coastal waters etc. Whereas man alters and manipulates these other processes, he has not chosen to modify active glaciers. Thus instead of dealing with all three components of the system, the process, the sediment, and the landform, only the last two are relevant in a discussion of applied glacial geology. Of course certain schemes, such as the towing of icebergs to water deficient sites and placement of lampblack on ice to accelerate its melting rate, have been mentioned . . . none have been prosecuted. Furthermore it is important to understand how the glacial process worked because such knowledge is crucial in predicting the occurrence, distribution, and type of feature that has resulted. Countless engineering failures have occurred and costly mistakes have been made when the results of the glacial process, the sediments and the landforms, were incorrectly interpreted. During Quaternary times glacier covered more than 30 per cent of the earth's land surface, and still cloak more than 10 per cent with ice that ranges to as much as 3 miles (5 km) thick. However, considerably more land was effected by glaciation because of outwash from the ice and loess blown hundreds of miles from the margin.

The cardinal rule in dealing with glacial deposits is the realization that no other type or group of sediments have as many different changes in such short distances both horizontally and vertically. This lack of site homogeneity must always be considered in programmes that call for their utilization. Such differences can occur in all sediment properties including texture, fabric, and composition. The mode of deposition, from till to ice contact to outwash, can also have abrupt changes and contacts. In similar manner no other sediment-based landforms can form by such a wide range of differing genetic processes, and be underlain by such contrasting materials. For example Kaye (1976) shows the engineering difficulties that occurred because the Beacon Hill end moraine had originally been interpreted as drumlins. This is a typical example of the law of uniformity. . . . similar appearing landforms are not all formed by the same process. The practical use of glacial geology knowledge can be applied in many fields that include engineering, mineral and water resources, land use planning, and law.

ENGINEERING APPLICATIONS

Although engineering covers a broad spectrum of man's endeavours, we will concentrate on the properties of glacial sediment as they influence engineering structures. Of primary concern is the correct assessment of the type and properties of the glacial materials. Because they can run the entire gamut of physical character, accurate mapping and borehole information is crucial to the success of any operation. Mollard (1973) presents a regional approach for analysis of important features, and Kreig and Reger (1976) show the technique that was used for aid in aligning the Alaska Pipeline. Of course not only must such work be done prior to construction, but the data must also be correctly interpreted. For example Legget (1974) discusses the building of the St. Lawrence waterway project and the millions of dollars that were lost by contractors who failed to interpret the drilling data accurately. The unifying theme of glacial deposits, is that they are similar in their lack of homogeneity. There are three large groups of glacial deposits, till, ice, contact, and outwash (see Glacial Geology, this volume).

Tills have often been considered by engineering generalists to be uniform in composition, physical properties, and even thickness. However, they show exceptional ranges in all these characteristics. Their geotechnical quality depends primarily on the four factors of (1) grain size distribution and mineralogy, (2) nature of the sequences of materials, (3) stress history during and after deposition, and (4) number, type, and orientation of joint planes. Of course many other factors are also involved in suitability of tills for use as fill and embankment materials, as substrate for structures, and for cuts to establish stable repose angles. These include water content and water table conditions, thickness and fabric of different units and other glacial deposits, and till type. For example lodgment till, and most glacial clays, are highly overconsolidated, flow till is only lightly overconsolidated, and ablation (melt-out) till is normal consolidated.

The engineering quality of tills ranges from compact and non-plastic, to weak plastic clayey tills. Their modulus of elasticity may vary from 150 to 7000 t/ft^2, and their undrained compressive strength from 1.5 to 30 t/ft^2. Silty or sandy tills can be compacted to a more dense state provided the soil has the proper water content at the time of compaction. A water content 2-3 per cent above the optimum of the soil is generally too soft for compaction. Degrees of compaction above 1.02 are only attained by substantial increase in compactive forces. Furthermore the compaction of till to higher densities does not appreciably change their compressibility and shear strength properties. Thus the compressibility of till is relatively independent of its initial density except for tills with unusually large amounts of plastic fines. Till texture also determines overall strength. A change in water content of only 2 per cent may change the remolded undrained strength by 50 per cent in clayey tills, whereas relative little change occurs for sandy tills. Great caution must be exercised when translating laboratory and field testing of till strength to operational situations. The in situ till mass is anisotropic with respect to deformational modulus so that many testing procedures overestimate the operational strength of the sediment. In situ coefficient of consolidation can run as high as 1000 times greater than the fill value from the same till. Table 9.1 shows a comparison of four different types of glacial sediment in terms of their suitability and behaviour for various engineering uses.

TABLE 9.1 : **Engineering and economic uses of glacial materials**

Type of Glacial Deposit	Bearing Capacity	Settlement	Slope Stability	Excavation	Use for fill	Use for Aggregate
Lodgment till	Best. Can contain irregular soft units. Silts subject to frost heave.	Smallest, but long term changes and differential displacement can occur.	Best with highest repose angles. However when wetted and dried subject to desication and movement along planes of discontinuity.	Very difficult. When wet hindered by plasticity. Can be nearly brick-hard when dry. Boulders a deterrent.	Especially suitable when impermeable materials are needed. May contain water content above optimum. Sensitive to moisture changes.	Unsuitable due to excessive fines.
Ablation till	Good but can be highly variable.	Slight to moderate	Moderate but lower cohesion than lodgment. Variable permeability can produce differential slope movement.	Moderately difficult. Boulders are special problem and composition and texture can provide wide variability in case of handling.	Good to very good. Silt particles can cause sensitive moisture changes. When selectively mined suitable for impermeable foundation.	Generally unsuitable due to excessive fines and variability. Processing required exceed economic merits.

Ice contact cohesion-less materials	Generally good, but interbedded till and openwork gravel produce variability.	Moderate. Most occurs during construction but long term consolidation possible and continuing changes due to till and clay in interbeds.	Generally at the angle of repose but instabilities can occur when gross textural sizes occur and with presence of till and clay.	Generally easy except in till and clay zones. Occasional boulders may present problems.	Generally good granular and free draining. Some caution necessary to prevent piping if materials are not blended.	Can provide important sand and gravel, but needs washing and screening. Kame deltas and kame terraces provide better source than kames and eskers.
Outwash cohesion-less materials	Generally good to very good.	Slight, but dependent on sedimentology of the deposit.	Invariably at the angle of repose	Easiest materials to remove and process.	Good granular materials that are free draining.	Best source of building and construction material. Sediments are best sorted, with less deleterious rock.

[Part of format after Fookes et al. (1975)]

Slope stability of tills is governed by many factors which include the bulk undrained and drained shear strength, consolidation or swelling properties, permeability and groundwater conditions, and degree of homogeneity and its interruptions. The repose angle for some England tills is determined by the presence of polygonal systems of ice-wedge clast relicts which create vertical planes of weakness. In British Columbia some dense tills have stability slopes less than 20^0 due to thin zones of highly plastic slickensided clay materials. Driving of well points to increase stability proved impractical because of the erratic character of interbeds.

Eden (1976) has showed the importance of water content of till in determining its construction suitability. Many problems were encountered in the cut and fill operation when constructing the alignment for the Quebec North Shore and Labrador Railway. The route traversed a ground moraine with drumlinized ridges for about 150 miles (240 km). In the cut sections the equipment became mired in sediment that liquefied, and nearly equal difficulties occurred in fill areas. The unstable till had a 14.3 per cent water content so that about 4 per cent of water content needed to be extruded to reach optimum dry density of 10.2. However due to the impermeable nature of till attempts at densification were initially unsuccessful because high pore water pressures were induced with attendant unstable behaviour of the sediment. Dikes were necessary in the construction of the Churchill Falls Power Project in 1969-70. Till was the major available borrow material but proved too unstable. It could not be sufficiently compacted to design density and permeability because the water content exceeded test specifications. Drier materials had to be transported from much greater distances thus increasing project costs.

Glaciolacustrine and glaciomarine sediments are notorious as produces of landslides and unstable terrain. Coates (1977), Dunn and Banino (1977), and Mollard (1977) discuss landslides in the extra-sensitive to quick clays that undergo liquefaction when disturbed, in such areas as New York, Norway, Alaska, and Canada. Such clays were responsible, for example, for most of the hundreds of millions of dollar damage done in Anchorage, Alaska during the 1964 Good Friday Earthquake.

Glacial aspects of dam siting and construction are discussed by Fluhr (1964), Legget (1974), and Philbrick (1976). Fluhr shows the importance of knowing the environmental setting as well as character of the materials for successful dam construction of four dams for New York City water supply in the Catskill Mountains. At the Downesville site till was used for the impervious core with 2-23 per cent clay, whereas kame deposits with glaciofluvial sands and gravel were used for the semipervious shell material with 4-7 per cent clay. The optimum density proved to be 129 lb/ft^3 at 9.3 per cent moisture content. The Portage Mountain Dam on the Peace River in British Columbia (Legget, 1974) is 600 ft high and 3,800 ft long. It is an earth-fill dam that contains 57.5 million yd^3 of glacial materials. Philbrick showed the importance of making a complete evaluation of all factors associated with glaciation in construction of the Kinzua Dam, Pennsylvania. Here site specific studies were made that led to the type of dam to be built, specifications for permeability of the foundation and embankments, source of aggregate, and foundation and slope problems associated with highway and railroad relocations. Unfortunately several of these factors were not analyzed in construction of the Lake Lee Dam near Lee, Massachusetts. The dam broke on March 24, 1968 killed two people and caused damages of $10 million. Cause of the failure can be attributed to piping which produced collapse after seepage was channellized under the structure. A similar catastrophe was averted in the Garrison Dam, North Dakota (Arnold, 1964) by construction of relief wells to prevent piping and development of sand boils.

ECONOMIC RESOURCES

Glaciation has produced and distributed a variety of products and features of economic value to man. Knowledge of the processes that have produced such benefits is often essential for their successful use and exploitation. Glacial deposits range from exotic occurrences of diamonds and gold to more prosaic sand and gravel resources. Since 1863 there have been 82 discoveries of diamonds in glacial drift of the Great Lakes region. From the early reports by Hobbs (1894) to the later work by Gunn (1968) many have attempted to trace such erratics to a Canadian "mother lode" but with no success. Gold in minute amounts occurs in many sand and gravel stream beds throughout the Great Lakes, transported there-from Canadian source deposits, and is especially prevalent in Indiana (Blatchley, 1902) although not occurring in sufficient abundance to be classed as an ore deposit. There are many theories that have been advanced to explain the lead-zinc district of Wisconsin, Illinois, and Iowa and one proposed mechanism for its origin involves meteoric waters generated by a glacial-artesian system (McGinnis, 1968). Such a theory holds that weight of the ice sheet distorts and displaces normal groundwater causing its flow to discharge sites where heat loss, pressure loss, and solubility changes resulting precipitation of the ore-bearing solution.

Analysis of till is being increasingly used as a method for ore exploration (Shilts, 1976 ; Alley and Slatt, 1976; Kujansuu, 1976), although certain principles were in use 240 years ago. The systematic geochemical mapping programme for ore prospecting in Finland started in 1971. These surveys include fabric analysis, petrographic analysis, trace-metal content granulometric analysis, and microfossil analysis. Results to date have differentiated five different tills, which will lead to the development of a new glacial chronology and dispersal directions of ice flow. In Newfoundland the probably source area of chalcopyrite-pyrite mineralization was determined on the basis of a float dispersion fan, concentrations of Cu, Fe, and S, clast lithologies, and fabric of till. Alley and Slatt (1976) point out the importance of studying the substrate by means of 6-8 ft deep pits, instead of placing complete reliance on surficial materials. Shilts (1976) provides guidelines for how a mineral, exploration of till should be conducted.

Glaciofluvial sand and gravel deposits constitute significant economic resources throughout the world. The large majority of sand and gravel in Scotland is of glacial origin, and such an origin constitutes 25 per cent of similar resources in England and Wales. In New York State 95 per cent of marketed sand and gravel comes from glacial deposits. Glacial sand and gravel occurs in ice-contact deposits such as kames, kame terraces, and eskers, and as outwash. The outwash deposits generally have highly quality materials because they are more uniform, have fewer deleterious sediments, and contain more resistant rocks. Due to the longer transport distances outwash materials are better sorted and stratified, contain harder rocks that have withstood comminution, possess less local rock and soil, and thus need less processing before use as a resource. Kame deltas can also serve as important aggregate sources. Proglacial lake beach environments may contribute valuable sand deposits when the duration of the lake was sufficiently long to produce winnowing action of sediments along the strandline. For example Glacial Lake Iroquiois beaches comprise significant sand deposits throughout the eastern Great Lakes. The search for appropriate sand and gravel resources can be shortened when there is knowledge of the glacial history of the site area. John Arborio, Inc. had the contract to build 30 miles (50 km) of the 4-lane Interstate 88 near Sydney, New York, and needed a sand and gravel source for the highway. This area is in the Susquehanna River valley, but as their consultant I knew that the local Catskill-type rocks had become incorporated into the glacial deposits and formed such a high percentage of the deposits that they would not pass State specifications for use due to large losses in the $MgSO_4$ soundness test. Thus important time and money were saved by being able to restrict the search for a suitable resource to other areas. In this region only sand and gravel in *through valleys* (those with communication through the Allegheny Escarpment) was of sufficiently high quality for use on Interstate roads; so it was in this glacial environment setting that an appropriate deposit was discovered. The reason for superior materials in the through valleys is because the sediments contain less local rocks, contain harder rocks transported from the northern part of the State, and have higher percentages of outwash-type sites.

Sand and gravel resources influence road construction in other ways. In the hilly glaciated Appalachian Plateau most government roads follow stream valleys where the majority of glacial stratified sediments are located. In construction of new highways private landowners displaced by the development may feel payment for their land is inadequate because of mineral resources. They initiate a lawsuit against the government in the search for a higher monetary award (see Coates,

1971, 1976). In such cases the geological testimony is crucial regarding the claim and assessment of the quality and quantity of the materials. Thus, it is important to establish whether the deposit has a uniqueness quality, will it pass specifications for use, and is there an available market. In some cases (Coates, 1971) it was important to map all sand gravel sources in a 10 mile radius to demonstrate that there was a great abundance of similar grade material in the Catskill region (310 million yd^3), thus showing that the material on the condemned property could not be claimed to be unusual or unique.

Glaciation has been instrumental in producing significant water resources in all glaciated regions. In North America glacial lakes contain 25 per cent of the total fresh water on the earth. The Great Lakes supply water for more than 25 million people and countless industries. Waterfalls created by glacial processes, such as Niagara Falls, are significant sites for development of hydroelectric power plants. Ground water in glacially-originated aquifers of sand and gravel constitutes a vital resource to numerous communities and industries. It is common that the glaciofluvial sands and gravels produced some of the highest yield wells in which pumping rates often exceed 1,000 gpm (gallons per minute). The ANSCO plant was located in the Binghamton, New York region because of the occurrence of abundant cold, clean, clear ground water in the glacial aquifers. Such cities as Endicott, Johnson City, Vestal, Elmira, and Corning in New York all derive municipal water supplies from glacial aquifers.

The siting of productive wells can be dependent upon information of the manner in which glacial processes operated in the area. For example it is important to know that in the valleys of southern New York and Northern Pennsylvania the strongest aquifers occur in valleys with south-flowing streams and the weakest aquifers are in valleys with north-flowing streams. The cause for the great discrepancy is that the glacial fill in south-flowing stream valleys contains higher percentages of sand and gravel. Thus when Frito Lay Industries decided to place their plant in a north-draining valley they were cautioned that at such a site they would be unable to develop a strong well system. Such advice was unheeded and thousands of dollars were wasted in an unsuccessful water drilling programme. However, C.J. Martin & Sons were able to drill successful wells at Whitney Point, New York when shown the proper site for exploration in a glacial moraine, and at Apalachin, New York when shown the best possibility for water in a kame and kettle complex. The water from Whitney Point was used for the cement in Inter-state 81, and at Apalachin the water was used for the Southern Tier Expressway.

The soil can be a valuable economic resource. Soils derived from glacial materials show an unusually wide range for crop productivity types, engineering capabilities, and homesite developments. On Long Island, New York those farmers who recognized the loess soil areas and established their fields at such locations obtained much higher production than others who had to farm the less fertile sand and gravel plains.

LAND USE

Glacial geologists can play a significant role in land use planning of glaciated terrain (LaFleur, 1974; Legget, 1974). For example sand and gravel resources should be mapped and placed into conservation zoning before urban sprawl development makes them unobtainable, as in much of the Chicago area. Soils and the character of the substrate should be classified into different use sequences. For example prior to purchase of an expensive lot a property owner should have knowledge whether the soil will pass the appropriate percolation tests and permit installation of a sewage system. In similar manner sites for sanitary landfills must be chosen with utmost care. Prior to the 1960s the location of landfills at abandoned gravel pits was standard practice in many communities. The substrate at such localities is often highly permeable, however, so that the resulting landfill leachate moves readily into the surrounding water supply systems. It is now recognized that landfills should be sited where the subsurface materials are impermeable and there is an available source of cover material that is also impermeable. Till is often the preferred type of deposit, so the character of the drift sheets, their thickness, and orientation is an important prerequisite in siting a successful sanitary landfill operation.

For larger developments and engineering projects the land use mapping should contain information on slope stability of materials, especially in permafrost terrain and in areas underlain by glaciolacustrine and glaciomarine strata. Even when prior information has been available, some land developers ignore it, and in some cases with disastrous results. The U.S. Geological Survey had correctly interpreted the nature of the Bootlegger Cove Clay but part of Anchorage, Alaska was built on it anyway. In somewhat similar fashion the Leda Clay, which is a notorious landslide producer throughout the St. Lawrence region, was known to occur at St. Jean Vianney, Canada. In spite of this a housing development was sited there and on May 4, 1971, 31 lives were lost when the material liquefied and produced landslide chasms to depths of 100 ft. Thus the cooperation of earth scientists who write the reports and the community planners is important in land use management strategies, such as occurred at Saskatoon, Canada (Christiansen, 1970).

Glaciation has affected land utilization in many other ways —some minor, and some major. Due to the till shadow influence on hills of southern New York, the thickness of glacial drift is often 8-10 times greater on south than on north slopes. In these uplands groundwater for wells occurs in bedrock, so that home owners who select sites on the south sides of hills must drill much deeper, and at greater expense, to obtain water than those on the north hillslopes. The grandeur of glacial scenery is unsurpassed on planet earth—the fiords of Norway, the Alps of Switzerland, Yosemite Valley, Niagara Falls, The Finger Lakes, etc. Such locales have become highly commercialized. The local people and the governmental bodies receive important financial rewards for investments in the tourist trade. Thus glaciers have produced land and scenic attractions which have led to land use developments, parks, wilderness, and exploitation. However, who can deny that although the entrepreneurs have benefited from such economic schemes, that those who partake of the wonders of nature are not also the richer in spirit for such communion?

REFERENCES

Alley, D.W. and Slatt, R.M., 1976: Drift prospecting and glacial Geology in the Sheffield lake—Indian pond area, northcentral Newfoundland, in *Glacial Till*, R.F. Legget, ed., Royal Soc. of Canada Special Pub. No. 12, pp. 249-266.

Arnold, A.B., 1964: Relief well on the Garrison dam and Snake river embankment, North, Dakota: *Geol. Soc. America, Eng. Geol Cast. Hist.* No. 5, pp. 45-52.

Blatchley, W.S., 1902: *Gold and diamonds in Indiana*: 27th Ann. Rept. of Ind. Dept. Geol. and Nat. Res., pp. 11-47.

Christiansen. E.A., (ed.), 1970: *Physical environment of Saskatoon*, Canada: Saskatchewan Research Council, NRC Pub. No. 11378, p. 68.

Coates, D.R., 1966: *Report on the geomorphology of the Cowanesque basin*, Pennsylvania: U.S. Army Corps of Engineers Cowanesque Reservoir Study, Baltimore, p. 27.

Coates, D.R., 1971: Legal and environmental case studies in applied geomorphology in *Environmental Geomorphology*, D.R. Coates, ed., Publ. in Geomorphology, State Univ. of New York, Binghamton, pp. 223-242.

Coates, D.R., 1976: Geomorphology in legal affairs of the Binghamton, New York metropolitan area, in *Urban Geomorphology*, D.R. Coates, ed., *Geol. Soc. America Special Paper* 174, pp. 111-148.

Coates, D.R., 1977: Landslide perspectives in Landslides, D.R. Coates, ed., *Geol. Soc. Amer. Reviews in Eng. Geol.*, Vol. III, pp. 3-28.

Dunn, J.R. and Banino, G.M., 1977: Problems with Lake Albany "clays" in Landslides, D.R. Coates, ed., *Geol. Soc. Amer. Reviews in Eng. Geol.*, Vol. III, pp. 133-136.

Eden, W.J., 1976: Construction difficulties with loose glacial tills on Labrador Plateau; in *Glacial Till*, R.F. Legget, ed., Royal Soc. of Canada Special Pub. No. 12, pp. 391-400.

Fluhr, T.W., 1964: Earth dams in glacial terrain, Catskill Mountain Region, New York in Engineering Geology Case Histories, No. 5, G. A. Kiersch, ed., *Geol. Soc. America*, pp. 15-29.

Fookes, P.G., Gordon, D.L., and Higginbottom, I.E., 1975: Glacial landforms, their deposits and engineering characteristics, in *The Engineering Behaviour of Glacial Materials*, Midland soil Mechanics and Foundation Engineering Society, Univ. of Birmingham, pp. 18-51.

Kujansuu, R., 1976: Glaciogeological surveys for ore-prospecting in northern Finland: *in Glacial Till*, R.F. Legget, ed., Royal Soc. Canada Spec. Publ. No. 12, pp. 225-239.

Gunn, C.B., 1968; Relevance of the great lakes discoveries to Canadian diamond prospecting: *Canadian Mining Jour.*, V. 89, No. 7, pp. 39-42.

Hobbs, W.H., 1894; On a recent diamond fin in Wisconsin and on the probable source of this and other Wisconsin diamonds *Am. Geologist*, V. 14, p. 31

Kaye, C.A., 1976: Beacon Hill end moraine, Boston: new explanation of an important urban features in *Urban Geomorphology*, D.R. Coates, ed., Geol. Soc. Amer. Special Paper 174, pp. 7-20.

Kreig, R.A. and Reger, R.D., 1976: Preconstruction terrain evaluation for the Trans-Alaska Pipeline Project, in *Geomorphology and Engineering*, D.R. Coates, ed. Dowden, Hutchinson & Ross, Stroudsburg, pp. 55-76.

LaFleur, R.G., 1974: Glacial geology in rural land use planning and zoning, in *Glacial Geomorphology*, D.R. Coates, ed., Publ. in Geomorphology, State Univ. New York, Binghamton, pp. 375-388.

Legget, R.F., 1974: Glacial landforms and civil engineering, in *Glacial Geomorphology*, D.R. Coates, ed., Publ. in Geomorphology; State Univ. New York, Binghamton, pp. 350-374.

McGinnis, L.D., 1968: Glaciation as a possible cause of mineral deposition, *Econ. Geol.*, V. 63, pp. 390-400.

Midland Soil Mechanics and Foundation Engineering Society, 1975: The Engineering Behaviour of Glacial Materials: *Proc. of Symposium April 1975*, Dept. Civil Engineering, Univ. Birmingham, pp. 272.

Mollard, J.D., 1973: *Landforms and Surface Materials of Canada: A Stereoscopic Atlas and Glossary*: J.D. Mollard, Regina, pp. 336.

Mollard, J.D., 1977: Regional landslide types in Canada, in Landslides, D.R. Coates, ed., *Geol. Soc. Amer. Reviews In Eng. Geol.* Vol. III, pp. 29-56.

Philbrick, S.S., 1976: Kinzua Dam and the glacial foreland, in *Geomorphology and Engineering*, D.R. Coates, ed.. Dowden, Hutchinson, & Ross, Stroudsburg, pp. 175-197.

Shilts, W.W., 1976: Glacial till and mineral exploration, in *Glacial Till*, R.F. Legget, ed., Royal Soc. Canada Spec. Publ. No. 12, pp. 205-224.

TEN

D. K. Ghosh

WEATHERING OF GRANITES AND DIFFERENTIATES AND THEIR INFLUENCE ON DAM FOUNDATIONS IN CENTRAL INDIA

INTRODUCTION

MANY river valley projects around Jhabua, parts of Bundelkhand, Korea and Patna states in Sarguja, Bilaspur uplands and Kanker in Chattisgarh, north and south Bastar, northeast Nagpur, northwest Bhandara and Chindwara etc. are located in the Archaean granites and differentiates. The author, while associated with geotechnical investigations of large number of such projects, has observed a characteristic topography, nature of bed rock and weathering profile. Sub-surface exploratory data from drill holes, trial pits, trench and road cuts besides canal and cut-off trench excavations for earthen/rockfill dams through the weathered granite material have supplemented useful information with regard to classification of weathered profile and studying their engineering properties for evaluation of dam foundations.

NATURE OF GRANITE-GRANITOID AND ROLE OF WEATHERING AGENT

The granite is predominantly coarse grained and prophyritic consisting of quartz, felspar, mica and some pyroxene. Its composition varies considerably with little or no sign of foliation. The granite, assimilating with the adjoining country rock during injection, looks like a hybrid material and has hitherto been referred to as 'granitoid' or differentiates. There are mainly two varieties of granite—one rich in alumina/potash felspar and some other characterised by richness of hornblende or pyroxene. The former variety 'biotitic or felspathic granite' shows pronounced weathering characters. The sodic and potash felspars decompose to kaolin and mica alters to clay. The granite is, in general, fairly compact and hard with 2-3 sets of master joints trending in NNE-SSW, ESE-WNW and NW-SE directions. Horizontal sheeting, parallel to ground surface, is also common especially on old relic plains. Opening of steep dipping joint surfaces and sheeting are considered primarily due to relief of confining pressures by uplift and unloading. Thus, decompression is supposed to be associated with the formation of internal cleavage planes in granite mass, which have facilitated main weathering agent—water to penetrate deep into the medium. Water has also caused granular disintegration on the surface giving a pitted appearance. The diameter of pits as seen in a number of outcrops around Sarguja and Kanker ranges up to 0.5 m. The ground water base is irregular and confined mainly to fracture porosity besides the weathering front.

TERRAIN CHARACTER

Physiographically, the granite terrain of central India occurs as scattered domes, cones and in a system of linear ridges rising above ground in small groups of pinnacled hills. These are associated with the process of exhumation (Ghosh, 1978). There are four types of slope in granite terrain—bare and covered rock dome, tor and inselberg, concave slope with free face and down slope in lower level and lastly a irregular profile (Plate 1).

On the basis of field evidences, the author considers two main types of granite topography—one as a relic of old igneous floor on which Dharwar and equivalent sediments were laid down and the other influenced by later intrusives into the Dharwarian metasediments and their equivalents (Fig. 10.1) In northern central India, the Bundelkhand granites and granitoids occupy a triangular tract as relic of old igneous floor. Similar types of granitic rock and topographic set-up also reappear around Jabalpur, Raigarh, some parts of Sarguja (in

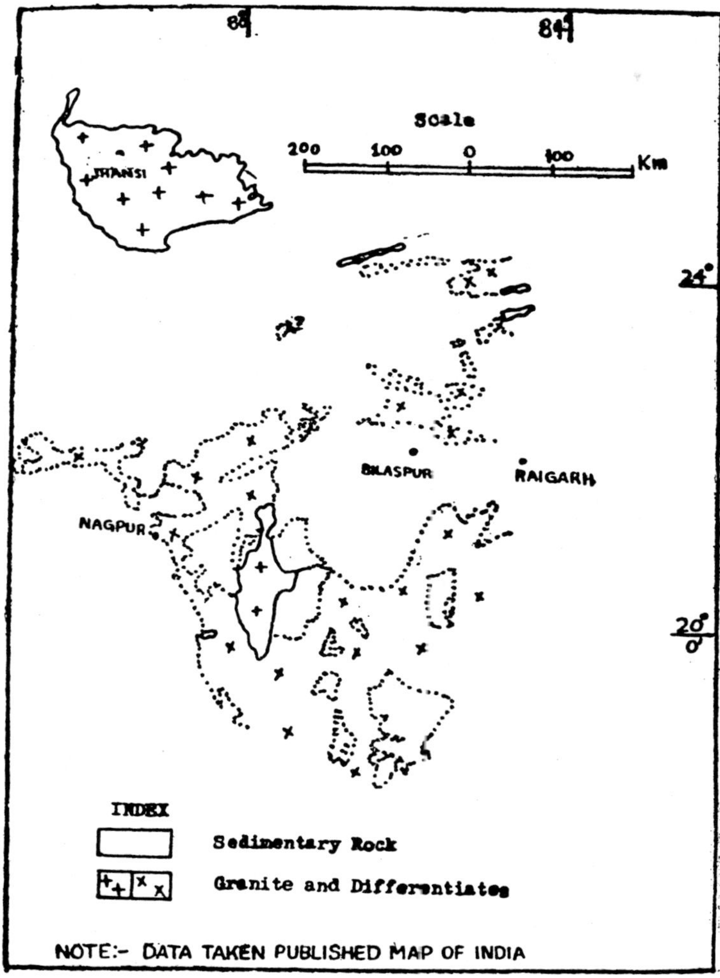

FIG. 10.1 : *Distribution of Granites and Differentiates of Central India.*

old Korea and Patna states), uplands bounding Bilaspur plains, Kanker-Siwaha areas in Chattisgarh, north Bastar, northeast Nagpur, Bhandara and Dongargarh, etc. In these areas, the granitic rock is generally coarse grained and forms scattered rounded hill covered with spheroidal boulders and broad plains which have shallow valleys in the overall gently undulating uplands. Pincer like drainage has formed arcuate landform, as seen in various parts of Sarguja and Kanker, probably due to unloading. The second category—gneisses and or granite-gneisses which are known to have intrusive relationship with the metasedimentaries, have produced a system of long narrow, deep carved ridges and valleys. Such outcrops are seen around Balaghat, Chindwara, Nagpur-Seoni and South Bastar, etc. However, there are some exceptions like the long narrow serrated ridges of quartz reefs in Bundelkhand landscape and the deep carved valleys in the granite-gneisses of Hasdo valley near Kathghora, Bilaspur (Plate 2). These are probably related to subsequent uplift of the basin floor.

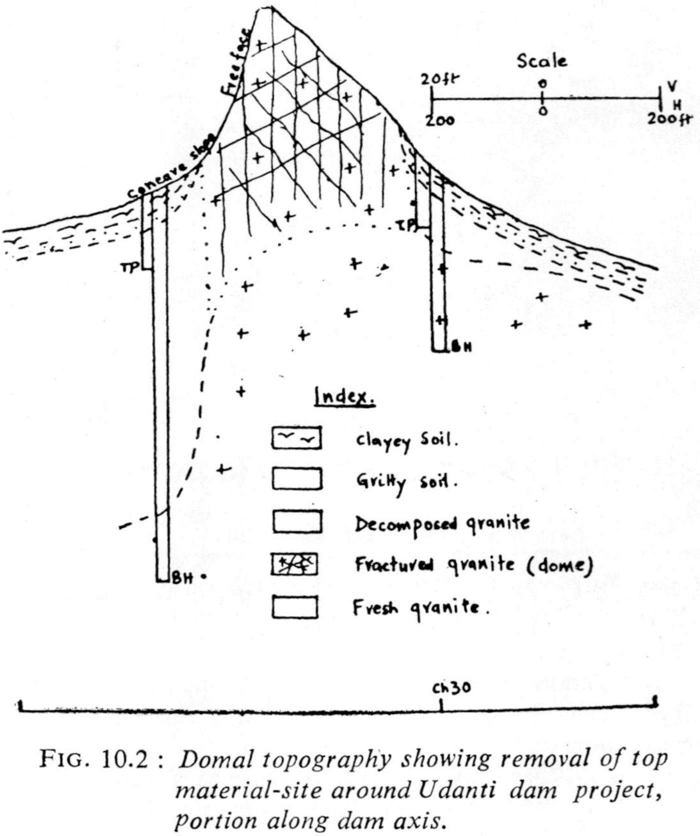

FIG. 10.2 : *Domal topography showing removal of top material-site around Udanti dam project, portion along dam axis.*

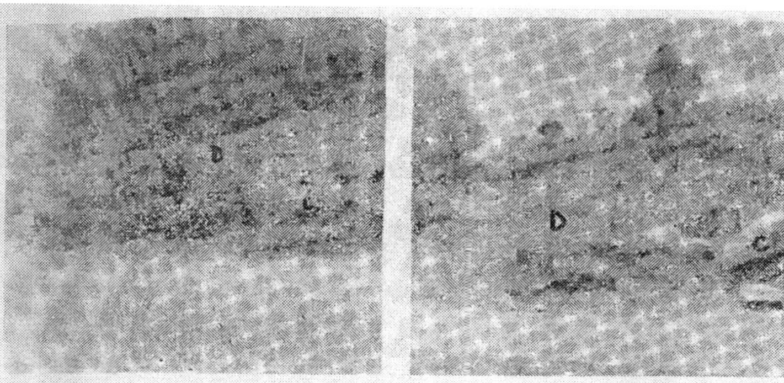

1. Showing a portion of covered rock dome slope, granite terrain near Ambikapur, M.P.

2. Showing deep carved valley in granite terrain at Hasdo-Bango dam site, M.P.

3. Showing moderately weathered granite (MW) overlain by decomposed ground mass with less than 50% crumbly patch and corestone, and top soil on the left bank of Hasdo-Bango dam project, M.P.

4. Showing fresh granite rack (PR) and decomposed matrix (D) with more than 50% of corestones (C), left bank of Hasdo-Bango dam project, M.P.

5. Upstream face of block 1 and key wall, left flank of Ghunghuta dam project, Ambikapuri; showing a granite weathering profile

S—Soil (sand and silt),

P—Weathered granite retaining some texture, though matrix is partially decomposed

6. Upstream face of excavated pit for block 1 and 2 showing weathering profile with kaolinised pockets (K) and partially decomposed crumbly patches (Cr) in completely weathered ground mass. Soil profile (D and S) is overlying the granite mass

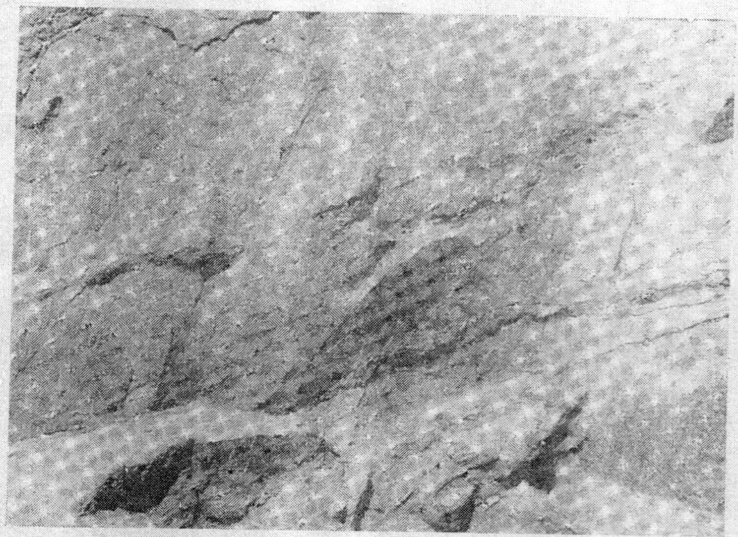

7. Showing fresh granite rock with initiation of weathering (W) along cracks (C), Left bank of Hasdo-Bango dam project, M.P.

8. Showing crumbly patches (Cr), Decomposed ground mass (D), and Soil (S) in a weathering profile of canal cut section of Banki dam project, Ambikapur

9. Showing junction of fresh rock (F), Decomposed rock with crumbly patches (P), decomposed rock without any patch of original rock, Ghunghuta project, Ambikapur

The present topography and disposition of streams in a dendritic pattern has no positive relationship with pockets of deep weathering, as it has been found in areas of relief and also peneplains. In general, deep weathering has given rise to negative relief and sculpturing of granite—accumulation of granite blocks, as seen around Dongargarh and some parts of Kanker/Sarguja etc., have shown areas of positive relief. In typical landforms of granite terrain, the domes rise upto approximately 60 m and the sheets range in thickness from 30 to 100 cm. (Fig. 10.2).

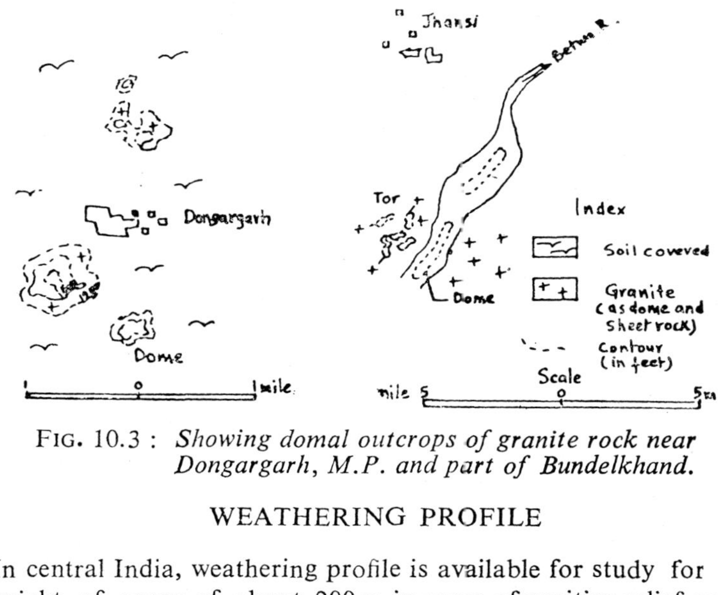

FIG. 10.3 : *Showing domal outcrops of granite rock near Dongargarh, M.P. and part of Bundelkhand.*

WEATHERING PROFILE

In central India, weathering profile is available for study for a height of range of about 200 m in areas of positive relief and 50 m in the zones of negative relief as noted through sub-surface exploration data of Sondur dam project in Siwaha-Kanker area and Ghunghuta dam project in Sarguja. On Hasdo-Bango dam project, a section of valley profile in weathered granite rock is seen for a height of about 80 m. In the topographic highs, the weathering profile is an elongate dome and the rock is mainly affected by disintegration as a result of mechanical weathering and top loose material, wherever not present, might have been removed during Talchir glacial environment of lower carboniferrous period (Fig. 10.3).

Topographic lows in the granitic terrain have shown decomposition of the rock mass essentially due to chemical weathering. The observed distribution of weathering stages tends to confirm that the granites and associates of Central India have been affected more by decomposition due to chemical weathering acting downwards from the land surface than granular disintegration. Hence it has been considered necessary to characterise the weathering profile by a weathering grade classification above the fresh rock. In all these cases, the effect and intensity of weathering was seen to increase towards the present ground surface in six stages shown in Table 10.1.

TABLE 10.1 : **Simple basic sequence of weathering in Central India on granite rock.**

Term	Grade	Description
(1)	(2)	(3)
Upper Regolith		Residual Soil: Pale yellow and red clayey/silty soil associated with organic material.
Lower Regolith		
Completely Weathered—Structureless Material	VI	Granite Soil: Intact, friable soil, pale yellow and red in colour, felspar and mica completely decomposed, to form clayey/silty soil which is weakly cemented and partially cohesive in nature. Complete granular disintegration due to combined action of chemical and mechanical weathering has produced at places gritty or sandy material—'grus'. It is non-cohesive and crumbles easily.
Highly Weathered	V	Partially structured rock material, which is decomposed and disintegrated to a weak intact groundmass having less than 50% of angular and subrounded cores tones, crumbly patches of unweathered granite rock. Original material fabric is lost in the ground mass (Plate 3).

TABLE 10.1—*contd.*

(1)	(2)	(3)
Moderately Weathered	IV	Partially structured rock material retaining original rock fabric where all mineral grains are not decomposed and having more than 50% of core-stones embedded in the groundmass (Plate 4).
Partially Weathered	III	Intensely fractured rock, completely stained to near 100% discolouration along discontinuity surfaces, rock fabric is intact; but weathered.
Slightly Weathered	II	Jointed rock with partial discolouration along separation planes indicating evidences of weathering.
Fresh Rock	I	Practically no visible sign of weathering.

It has also been noted that a complete weathering profile in six stages (Table 10.1) is not always seen in progressive order and the sequence, in Table 10.2, shows that the pattern of weathering at times may also be complex due to overlapping and/or inter-relation of weathering grades. In such situation; the rock : soil ratio and the presence or absence of original rock texture have been found to be useful tools for practical assessment of weathering grade classification. On the basis of this assessment only four possible grades were discernible—all rock, rock and soil with original texture, rock and soil without original texture and all soil. Further, the depth of weathering has gone down to ground water base which coincides with the weathering front or fresh rock level.

TABLE 10.2 : **Showing different pattern of weathering in some sites in Central India**

	Depth	Grade
Ghunghuta dam site, Ambikapur Sarguja (Plate 5, 6)		
Organic Soil	0.25 to 0.5 m	
Residual granite soil, silty and sandy in nature.	up to 10 m	VI
Partially weathered granite	up to 5 m	III

TABLE 10.2—*contd.*

	Depth	Grade
Fresh granite to slightly weathered granite.		II-I
Hasdo-Bango dam site, near Kathghora, Bilaspur (Plate 3, 4)		
Organic Soil	up to 1.5 m	
Granite Soil	,, 3.2 m	VI
Highly weathered granite	,, 25 m	V
Moderately weathered granite	,, 15 m	IV
Partially weathered granite	,, 5 m	III
Slightly weathered granite	,, 15 m	II
Sondur dam site, near Siwaha/Kanker, Raipur.		
Organic Soil	up to 2.5 m	
Granite Soil, clayey and silty	,, 4.0 m	VI
Partially structured granite retaining some texture in decomposed matrix—silt/grit with pockets of grus.	,, 3.0 m	V—VI
Structured granite as sub-rounded angular corestones and crumbly patches in decomposed matrix silty and sandy in nature.	,, 17 m	IV
Slightly weathered granite with intense fracturing and jointing Fresh granite.	,, 27 m	III
Udanti dam site, near Gariaband, Raipur.		
Organic Soil	up to 1.5 m	
Gritty/Silty Soil matrix with less than 50% crumbly patches and corestones of granite rock.		V
Weathered granite corestones and crumbly patches in less than 50% decomposed matrix.	,, 20 m	IV
Unweathered granite rock with spaced joints.		I

Note : The cut-off trench for earthen dam across Sondur has also indicated highly variable weathering profile with the presence of hard silty soil immediately above decomposed granite of grade IV and also grus and gritty soil above granite of grade III and IV in some portions.

It has been possible to visually recognise the various changes in the weathering profile and it has been observed that the style and rate of weathering is by and large controlled by porosity and permeability of the granite rock. Initial weathering process might be appreciably slow (Plate 7) along joints and cleavages which isolates the jointed blocks to form corestones. Further ingress of water along minute cracks in rockmass and cleavage planes of felspar and mica lead to rapid attrition and alteration of such mineral constituents to form crumbly patches (Plate 8). Weathering crusts on the surface holds water and aids further alteration.

Typical granite weathering seen on spurs and in deep glens has shown that deep pockets of regolith are the areas of valley formation where minor streams cut through the regolith to expose some of the ribs of unweathered granite rock (Plate 9). During course of erosion, the basal surface of weathering is exhumed and the corestones left behind as soft material of regolith is washed away.

ENGINEERING PROPERTIES OF WEATHERED GRANITE AND INFLUENCE ON DAM FOUNDATION

Weathered granite has been encountered in many civil engineering projects and consequently its engineering properties and behaviour have been studied. A fairly well elaborated description of weathering profile and their classification have provided a basis for presenting an engineering grade classification (Table 10.3) for use of weathered granite in the dam foundation and as construction material.

Samples collected from various horizons of the weathered profile have been tested in the laboratory for determining their engineering properties (Table 10.4) and studying their behaviour for foundations.

The marked increase in wet density and moisture content reflect progressive opening of grain boundaries, development of microcracks. The alteration of felspar and mica has increased the clay content in the soil which resulted in the decrease of the permeability values. Though upper horizons of the soil is generally dry but the lower levels (below 3—5 m) are wet. In

TABLE 10.3 : Engineering Grade Classification of Granite Country in Central India

Grade	Degree of Weathering	Field Recognition	Engineering Property
VI	Granite Soil	The rock is completely changed to clayey, silty, gritty soil with no fabric of parent rock.	May erode easily unless a hard cap is present. Requires selection before use as fill. High cuts unstable. Gritty layers are constant source of leakage. Unsuitable for large structure but can form foundation for low earth dam.
V	Highly Weathered	Pulverised to compact soil retaining no original rock texture with embedded corestones and crumbly patches.	Suitable for low earth dams but unreliable for concrete/masonry and large dams due to errotic boulders, can be excavated by ripping, but high cuts unstable. Requires erosion portection measures.
IV	Moderately Weathered	Pulverised soil with some rock texture and embedded corestones and crumbly patches.	Grade IV is comparatively better material for low earth foundation.
III—II	Partially Weathered	Fractured granite but highly stained.	Permeable through joints, requires explosives, stained rocks are doubtful aggregates. Can be rendered suitable by grouting for low masonry/concrete dams and medium earth/rock-fill dams.
I	Fresh Rock	Less fractured granite.	Suitable for all civil engineering structures and aggregates, requires blasting.

TABLE 10.4 : **Typical Range of Engineering Properties of Weathered Granite**

Type of Material	Bulk Density		Moisture %	Fineness Modulus	Clay %	K-Value (Lugeon)	Other
	Dry	Wet					
Granite Soil—VI (Clayey/Silty)	1.49—1.54	1.87—1.90	20.74	3.27	30.2	3	—
Grade V	1.58—1.75	1.90—2.03	19.6—22	2.5—3.1	18.6—38.4	3—5	—
Grade IV	1.66	—	—	—	20.2—37.5	8—39	—
Grade III—II	1.76—1.80	—	—	—	—	16—25	'RQD' 25—50% for top 5 m.

situ condition, the soil appears cohesive and intact which otherwise loses inherent strength and bondage on optimum saturation, thus crumbling easily. The deformation behaviour of the decomposed granite material could not be interpreted successfully on the basis of available characteristic weathering profile at three different sites though the visual examination indicated a fair similarity. Hence it has been decided to examine the suitability of decomposed granite for use in the dam foundation on the basis of prevailing physical and engineering properties, and also in view of the fact that the decomposed granite material with high clay and silt content did not show any appreciable consolidation on optimum saturation in confined conditions. Satisfying the design parameters for controlling subsurface seepage and erosion through such type of foundation material, it has been found that the granite soil and decomposed material with clay/silt content of more than 30 per cent, K-values within 5 lugeon are suitable for the stability of low to medium earth/rockfill dams and it is not always necessary to take the foundations to rock level. Such decisions have reduced the depth of cut-off trench excavation and cost of construction in the 28 m high earthen dam under construction across Sondur near Siwaha region of Madhya Pradesh. In view of the heterogeneity of the foundation material at places, it has also been decided to provide relief holes at the downstream of dam to control exit gradient of the seepage flow. The decomposed granite material of grade IV and V has, however, shown better physical and engineering properties, barring some erratic values of permeability. Test grouting in this medium has indicated non-acceptance of grout in most of the sites but partial acceptance of grout up to 3 bags of cement in a ratio of 1 : 20 and 1 : 10 with water has also been recorded. In such cases the grout travel is not more than 1.5 m. Under these circumstances, it was not possible to uniformly improve the strength character of the foundation by grouting and as the nature of grout travel was erratic due to heterogeneous material character.

The partially weathered granite of grade II and III could successfully be grouted and then considered suitable for low to medium high concrete/masonry dams. In this type of material, the rock quality designation (RQD) of less than 40%

in the top 5 m depth has shown a poor quality rock which required extensive consolidation treatment for improving the strength character. On Hasdo-Bango dam project near Kathghora, the weathered granite profile on the left bank has necessitated changes in the type of dam from a masonry structure in blocks 1 to 9 to a rockfill structure for a height of about 20 m. At this site the depth of weathering to fresh rock is 25 m.

Stained granite rock has not been found suitable as construction aggregates but the decomposed material to silt and clay could selectively be used as fill material. The most undesirable part in weathered profile is the 'grus' or granite sand portions which have shown permeability values of the order of 80-100 lugeon and more. It is a cohesionless and low shear material like loose sand and needed to be removed from the foundation of earth/rockfill dams, for controlling underseepage and erosion.

CONCLUSIONS

Topography, nature of granite and weathering profile supplemented by surface data have indicated a sequence of weathering above fresh rock as intensely fractured rock, structured bed rock with corestones, decomposed bed rock matrix with or without corestones and structureless material as granite soil. The overall pattern of weathering is erratic and deep to a depth of about 50 m. Decomposition and further disintegration have formed clayey/silty/gritty soil when the weathering process is complete and partial decomposition is characterised by the presence of crumbly and corestone blocks, in the pulverised ground mass. Coarse granite has formed in situ 'grus' due to granular disintegration.

Predominant foundation problems in the weathered zone include internal erosion and resultant differential settlement, if any, and non/partial-acceptance of grout in the heterogeneous matrix. Low RQD in the intense jointed rock above fresh rock level requires special treatment for consolidation of foundation. An engineering grade classification has been presented for use of weathered zone in various civil engineering structures.

Laboratory tests are required for determining shear parameters for each type of weathered material to examine fundamental relationship between engineering properties and weathered matrix.

SUGGESTED READINGS

Cooke, R.U. & Doornkamp, J.C.: *Geomorphology in environmental management*, Clarendon Press, Oxford.

Ghosh, D.K., 1978 : Role of geomorphology in project planning of Central India, *Symp. Morphology and Evolution of Landforms*, New Delhi, 78.

Ghosh, D.K., 1977, 78, 79 : Geological reports on the Sondur dam project, Raipur M.P. (Unpublished GSI report)

Ghosh, D.K., 1978 : Geological report on the Udanti dam project, M.P. (Unpublished GSI report)

Ghosh, D.K., 1979 : Geological report on the Ghunghuta dam project, M.P. (Unpublished GSI report)

Ghosh, D.K., 1978, 79 : Geological report on the Hasdo-Bango dam project, M.P. (Unpublished GSI report)

ELEVEN

N. K. De

LAND CLASSIFICATION—A STUDY IN APPLIED GEOMORPHOLOGY

INTRODUCTION

LAND CLASSIFICATION implies the development of different kinds of lands into defined categories, according to the characteristics of the land. There are different ways of classifying land for different purposes by different scientists. As there are many ways of approaching the problem it is not easy just how to begin and carry out the land classification work.

To attempt to develop land without first assessing resource potentialities and planning their development is to invite economic failure and permanent impoverishment of natural resources. Before development begins, therefore, the nature of the resource potentialities should be carefully studied and an assessment be made of the suitability of the land for various kinds of use. This objective can be rapidly attained only by a concentration of efforts on the planning of development and on the implementation of the plans after initial classification of land as inventory of natural resources based on geomorphological attributes which are directly related to their ecological set-up. Thus geomorphology, the science of earth forms has important applications in the investigation, evaluation and classification of land as a resource. As the nature of the ground surface is an important indicator of the character and distribution of rocks, minerals, soils, vegetation, climate, drainage etc., the application of geomorphological knowledge in land classification becomes an essential tool.

Study Area

The area studied covers about 78 square kilometres in northern Calabria, South Italy. The area is approximately bounded by the lines of latitude 39° 35′ N, 39° 39′ N and longitude 16° 17′ E and 16° 24′ E. It is covered by a single aerial photograph, the average scale of which is 1 : 35000.

In the present work an attempt has been made to collect a detailed information of large areas as quickly as possible and to classify, map and describe areas based on the land potential characteristics. The importance of geomorphological attributes been and their significance in other studies have taken into consideration as the basis of land classification. Here the initial classification of available land in terms of physical factors as a basis for subsequent socioeconomic classification has been clearly recognised.

Methodology

The methodology followed in this study was similar to that of the surveys carried out by the commonwealth scientific and Industrial Research Organisation (C.S.I.R.O.), Division of Land Research and Regional Survey, Australia, with modifications. The method of survey is based on the assumption that each type of land surface or land type is represented on aerial photograph by a distinctive photo pattern. A fundamental assumption behind the concept is that each part of the land surface as expressed in the photo pattern is the end product of an evolution governed by the parent geological material, geomorphological processes, past and present climates and time. Thus an area or group of areas throughout which there is a recurring pattern of topography, soils, and vegetation can be recognised as a land complex or land type which is essentially consists of one or more land units (component elements of land complex or land type). Thus a distinctive photo pattern is essentially a distinctive land complex or land type.

The methods used in the survey can be sub-divided into the following :

(i) Pre-field methods.
(ii) Field work methods.
(iii) Post-field methods.

PRE-FIELD METHODS

The essential requirements of the pre-field work are (a) a complete cover of aerial photographs; (b) literatures, reports and maps available for the area concerned. The average scale of the aerial photographs used in the study was 1 : 35000 (approximately).

The main objective is the systematic preparation for field work. This includes gaining knowledge of the area in which the survey is to be carried out from existing reports and maps; the study of the aerial photographs; and in drawing up of a system of traverses and localities for detailed sampling to embrace all types of country to be found within the area. This stage of work includes the following phases :

(i) The collation and interpretation of the available literature and maps relating to the survey area are made.
(ii) The aerial photographs are studied using mirror stereoscope.
(iii) Traverse routes are carefully planned and marked on the photograph.
(iv) The major photo patterns corresponding with the major land complexes/land types are demarcated and classified.
(v) A preliminary appraisal of the main land complexes/land types and their component units or constituent elements is made from the air photo patterns.
(vi) Sample areas close to traverse routes are carefully and judiciously selected in each of the land complexes/land types according to photo patterns.
(vii) The itineraries of traverses for each day are planned.
(viii) Localities for base camps are selected.

FIELD WORK METHODS

In this type of survey, the survey party in the field must be well-equipped and must at least consist of three scientific personnel (one geomorphologist, one pedologist and one ecologist) who individually will be responsible for their respective field observations and will keep records of all of their observations whilst traversing, as well as of the sample areas.

Work during field work can be sub-divided into three parts :

A. Work during traversing.
B. Work in the Sample areas.
C. Work after sampling.

Work during traversing consists of the following :

(i) To have a general idea of the whole area and the land complexes/land types a general traversing is made.
(ii) Short descriptive symbols are then marked (A, B, C etc.) on the photograph for each of the land complexes/land types.
(iii) Sample areas are finally selected in all the land complexes. This is achieved during the traverse and/or after the traverse.
(iv) During traversing relevant field observations are made and land complex/land type boundaries are marked on the photograph.

Work in the sample areas consists of the following :

(i) Having arrived at the sample area, a quick reconnaissance of the form, material and vegetation is made.
(ii) If necessary the boundary of the sample area is changed in the light of the field appraisal.
(iii) Landform as seen in the field is correlated with the air photo patterns. For this pocket stereoscope is used. Photo patterns are thus interpreted.
(iv) Finally, the units to be sampled are discussed and agreed upon.
(v) The geomorphologists directs the other field scientists in their systematic examination and recording of soil and vegetation characteristics of the units.
(vi) The sample area as a whole and its component units are then studied. This includes mapping, the preparation of a block diagram showing all the component

units, the study of forms and their measurements and the study of the internal characteristics and external relationships of the units concerned including their extent, characteristic slope and its direction, study of surface materials and their characteristics, the character of the ground surface, the study of micro relief, local relief and the study of soils and vegetation.

Work after sampling can be summarised as follows :

(i) The relationships of the units are discussed and determined by the scientists.
(ii) For each sample site the land form, soil and vegetation information are then incorporated in tabular form on the sample area sheet prepared for the purpose.
(iii) A true slope profile and/or block diagram for each of the sample areas is drawn showing all the units and is incorporated in the sample area sheet.
(iv) Finally, land complex/land type symbols are decided upon.
(v) The sample area numbers and the land complex/land type symbols are marked on the sample area sheet as well as on the photograph.
(vi) All the units sampled are marked on the photograph for each sample area.

POST-FIELD METHODS

This stage of the work is carried out in the laboratory as well as in the base camp and consists of the following phases :

(i) The sample area sheets are grouped into land complexes/land types.
(ii) An idealised block diagram is finally drawn for each land complex/land type showing all the units and their spatial relationships as compounded from the various sample areas.
(iii) General description of the land complexes/land types are made.

(iv) The land complex/land type boundaries along the traverse routes are re-examined in the light of the experience already gained and data collected.

(v) The land complex/land type boundaries are extrapolated beyond the traverse routes to cover the whole of the area concerned on the basis of field experience, available data and photo interpretation.

(vi) New land complexes/land types (not sampled) are recognised and/or grouping of land complexes (if necessary) is made during extrapolation. These are the final land complexes/land types which constitute the mapping types to be recognised and described.

(vii) The descriptions of land complexes/land types (mapping types) ultimately recognised are finalised.

(viii) The land complexes/land types finally mapped are compared in detail.

(ix) Finally, the land complexes/land types are defined.

Land Complexes/Land Types

Six land complexes/land types have been mapped and described in the survey area (Fig. 11.1). The land complexes/land types and their component units are described in tabular forms and illustrated by block diagrams or by sketch plans. The relief of each of the block diagrams is either exaggerated or reduced in varying amount and no attempt is made to provide vertical or horizontal scales or to depict true relative areas of units. The measurement of units in each land complex/land

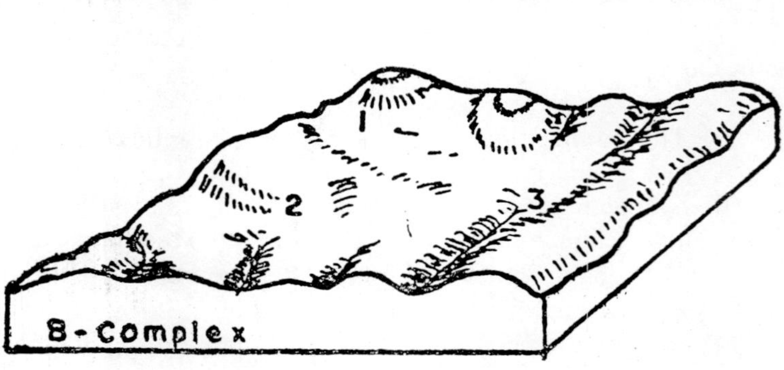

FIG. 11.1. *B-Land Complex/Land Type*

type are based on random sampling on air photos. They are merely an approximate guide and relate only to the sample areas.

Out of the six land complexes/land types (A, B, C, D, E and F) identified in the area only B-land complex which covers only about 20 sq. km. has been described in tabular form with a block diagram. In the same way the remaining six land complexes/land types can be described.

General	:	Undulating low hills on Tertiary Sediments, local relief about 12 m.
Geology	:	Tertiary calcareous marks.
Geomorphology	:	Undulating low hills characterised by smooth, almost flat but slightly rounded crests, gentle valley sides and incipient, dry sub-parallel head water tributary floors.
Climate	:	Humid mesothermal subtropical climate—CS type.
Soils	:	Very deep, Olive soils of fine texture, pH > 8.0.
Vegetation	:	Cultivated with wheat, barley and mustard.

Units	Landforms	Soils	Vegetation
1.	*Crest* : Extensive, smooth rounded with surface cracks, slope 2^0—3^0, about 60 m wide.	Very deep, pale Olive to Olive grey brown soils of fine texture, pH > 8.00.	Cultivated : Wheat, barley mustard.
2.	*Valley Slopes* : Rectilinear with surface cracks; slope 10^0—15^0 about 60 m long.	Very deep, Olive grey to grey brown soils of fine texture; pH > 8.00.	Cultivated : Wheat, barley, mustard.
3.	*Tributary headwater Zone* : Concave, broad shallow valleys dissected by few runnels 45 cm (wide) × 30 cm (deep); cracking surface; longitudinal gradient 4^0, about 30—40 m wide.	Deep, olive brown soils of fine texture; pH > 8.00.	Cultivated : Wheat, barley, mustard.

Land Classification

The area has been divided and mapped into six land complexes/ land types which are nothing but the initial classification of land based on geomorphological characteristics, soils and vegetation taking into account their interaction between them.

The land complexes/land types (Fig. 11.2) are further classified into two broad categories (a) erosional and (b) depositional, according to the history of the land surface and relief. In this way, in the case of the erosional land

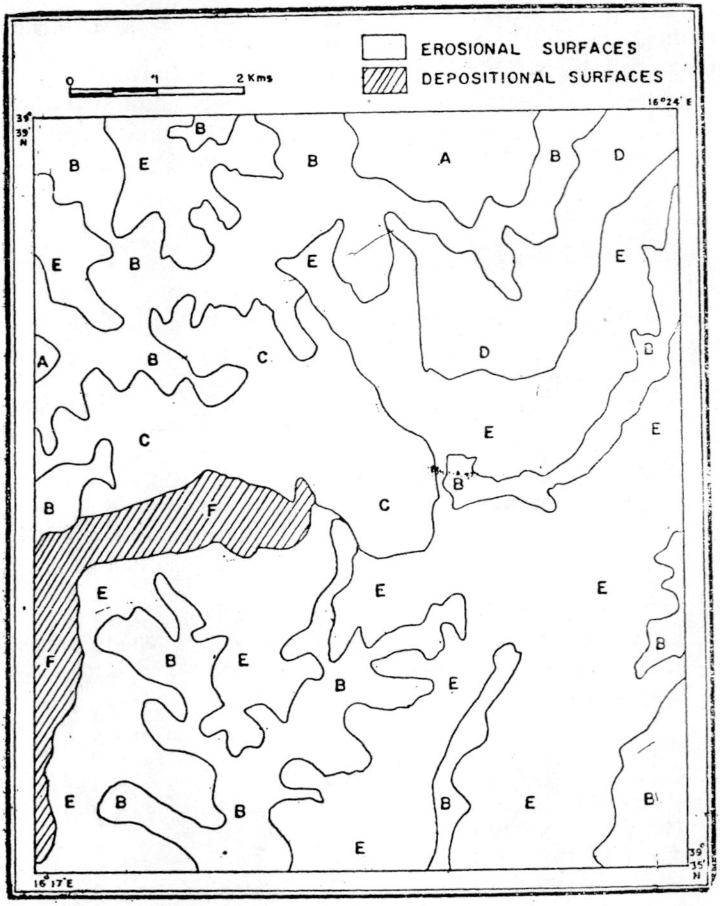

FIG. 11.2 : *Land Complexes*

surfaces, the inherited as well as the younger soil landscapes are explained. The depositional land surfaces can again be classified according to the nature of the deposits. Thus major contrasts in landforms, soils and vegetation are emphasized.

Discussion

There are many kinds of approaches for sub-dividing a country. It is important to note that in any classification, observable and recordable features must be taken into account. Thus the initial classification of available land in terms of physical factors should be the basis for future economic planning. Landform which is the end product of an evolution governed by the parent geological material, climate, soils, vegetation, drainage and the geomorphological processes operating through time plays an important role in this respect. The methodology given in this paper aim at fulfilling these requirements. However, any standard method may not be suitable for all purposes. Modifications are necessary to suit the requirements of different areas and people.

It was not possible in the time available to map the areas in detail. Therefore, a method was sought whereby the field details can be mapped as complexes of the country. The factual land complex/land type descriptions provide a basis for assessing land potentialities. The areas covered by each land complex/land type can be measured from the map and their relative importance can be judged. Thus the mapping of land types or complexes will indicate those parts of the region which warrant the most attention and those which might be left for later examination. Again as the survey is scientific in nature, the results establish a foundation on which future more detailed work or special scientific investigations might be based.

The technique assumes that the land complexes/land types and their component units are flexible and can be adjusted to the degree of complexity at any scale of working. Finally, the use of air photographs and the new techniques of photo interpretation have facilitated the survey techniques. It facilitates minimum sampling and allows the extrapolation of the recorded data from small areas beyond the traverse routes to cover large areas.

CONCLUSION

The identification and recognition of the land units and their tabular description enables a complex region to be analysed into all its component areas of different character or potentiality. These units are flexible and can be adjusted to the degree of complexity at any scale of working as stated earlier.

The map produced by this survey can be used in a variety of ways by other authorities and can serve as a very satisfactory framework for systematic sampling and correlation especially in geographic and economic studies or in other words the natural divisions of landscape provide a common basis for a wide range of studies. Thus it can be said that the methodology used in this study not only help assessing land potentialities, classify land, map and describe large areas of which little or no information are available but also it can safely and satisfactorily be applied to all kinds of ecological studies including soil surveys, vegetation surveys, geological surveys, hydrological surveys, anthropological surveys and land used surveys.

SUGGESTED READINGS

Agricultural Experimental Station, 1940: *Proc. of the First National Conference on land classification Bull.* 421, Columbia, Missouri.

Albeiter, J.K., 1940: Productivity rating of soil types, *Proc. First Nat. Conf. on land classification, Bull.* 421, Columbia, Missouri.

Azzi, G., 1956: *Agricultural Ecology*, Constable & Co., London.

Bhattacharyya, S., 1975: *Productivity study and land classification of Khargram-Indrani region*, Dist. Murshidabad, W. Bengal, Ph. D. Thesis, C.U.

Conrey, G.W. 1940: Primary soil features considered in land classification, *Proc. First Nat. Land Classification, Bull.* 421, Columbia, Missouri.

Economic Planning Agency, 1968: Fundamental land classification survey, *Koriyama*, National Land Survey, Japan.

Gerassimov, I.P. et al., 1958: Three scientific schools dealing with the general question on classification and their natural connections, *Geog. Rev. India*, Calcutta, Vol. 20., No. 1—4.

Hills, G.A. 1949: The classification of Northern Ontario Lands according to their potential for agricultural production soil Report I, Ontario Deptt. of Lands & Forests, Toronto, Canada.

Hudson, G.D., 1940: The contributions of geography to land classification, *Missouri Agri. Expt. St. Bull.* 421.

Jacks, G.V., 1946: Land classification for land use planning, *Imperial Bureau of Soil Science*, England, Tech. Comm. No. 43.

Kellogg, C.E., 1936: Development and significance of great soil groups of the United States, *U.S.D.A.*, Misc. Pub. 229.

Kellogg, C.E., 1938: Recent trends in soil classification, *Soil Sci. Soc. Amer. Proc.* No. 3.

Kellogg, C.E., 1940: The contributions of Soil Sci. and Agronomy to Rural Land classification *Proc. First. Nat. Conf. on Land Classification, Bull.* 421, Columbia Missouri.

Land Classification in the U.S. 1941: Report of the Land Committee to the National Resources Planning Board, Washington.

Linton, D.L. 1951: The delimitation of Morphological Regions, London, *Essays in Geog.* (L.D. Stamp & S.W. Woolridge eds.), London.

Matsui & Iwasa, 1967: The classification of Paddy Soil around the valley along the mid and downstream of the river Arakawa on the north of Tokyo, Misc. Reports, *Research Inst. for Natural Resource*, No. 68, Japan.

Panse, V.G., et al., 1954: Statistical Methods for agricultural Workers, *Indian Council of Agri. Research*, New Delhi.

Ramaun, E., 1918: Bodenbildung and Bodeneinteilung, (system der Boden), Berlin.

Saha, P.K., 1969: *Rating of soil productivity and land classification in the Dwarka-Brahmani catchment area*, District Birbhum, Ph. D. Thesis, Calcutta University.

Singo Mitsui, 1955: Inorganic nutrition fertilization and soil amelioration for lowland rice, Yokendo Ltd., Tokyo.

Storie, R.E., 1933: An index of rating of agricultural value of soils Calif. *Agri. Expt. Sta. Bull.* 556.

Storie, R.E., 1950: Transact, Fourth Inter. *Congr. Soil Sci.*

Takamasa Nakano, et al., 1966: Recent progress in geographical cartography and photo interpretation for area studied and land classification of Japan, *Japanese Geog. Spl. Publ.* No. 1.

Vileneskie, D.G. 1963: *Soil Science* (Translated from Russian) Israel Programme for Scientific translation, Jerusalem.

West Midland Planning Group, 1947: *Land classification in West Midland Region*, London.

Z. Vorykin, K.V., 1963: Scientific Principles for Agro-Production Classification of Lands, *Soviet Geog. Rev. Trans.*, Vol. 4.

TWELVE

Mieczyslaw Klimaszewski

A DETAILED GEOMORPHOLOGICAL MAP

THE RELIEF--AN IMPORTANT ELEMENT OF THE GEOGRAPHICAL ENVIRONMENT

EVERY society lives in a certain territory, under the conditions of a determined geographical environment. That environment forms the base of man's life and activity on the earth; therefore, the knowledge of the geographical environment and its particular elements is necessary to master and utilize it rationally. Besides the geological structure, soil cover, climate, water, and plant cover, the relief of the earth surface is a specially important element of the environment. These particular elements have been studied for a long time, and their spatial differentiation has been represented on geological (since 1850), palaeological, climatological, hydrographical and phytogeographical maps of general scales. Information about the configuration of the earth surface is presented on the topographic and not on geomorphological maps.

THE STUDIES ON THE RELIEF—ACHIEVEMENTS AND DEFICIENCIES

Topographic maps have since oldest times informed about the main features of the configuration of the earth surface; at first by means of the mound method (mountainous areas), since the XVIII c. by means of the chachure method, and since the XIX c. by means of the contour lines method. The analysis of these maps allows to distinguish and define morphographic and morphometric features of the relief forms represented by means of contour lines, as well as to interpret genetically the larger forms. On the other hand, topographic maps do not inform either about the origin or the distribution of smaller and less pronounced denudative forms (e.g. landslides, monadnocks, debris mounds), fluvial forms (e.g. small, but very important fragments of river terraces, rocky and alluvial beds), karstic forms (e.g. funnels), glacial forms (e.g. röche; moutonnées), nival melting, aeolian and riparian forms thus, in fact these maps fail to inform about many factors and processes to which these forms owe their origin. Neither do they inform about the age of these forms and their relation to the geological structure which renders impossible the reconstruction of the development of the relief on the basis of an analysis of the topographic maps, and the prognostication of their further developmental tendencies. Thus, a topographic map does not—and cannot—replace a geomorphological map, even though it has played that role over 100 years checking the progress in geomorphology. On the other hand, a topographic map prepared by the representatives of geodesy (topographers) gives a general picture of the relief and may be considered a useful basis for further studies, especially for geomorphological mapping. Unfortunately, for a long time, the studies did not comprise all forms but were limited only to some chosen, very characteristic, or even "fashionable" ones. Thus, river valleys, or rather some of their stretches were investigated, and above all the terraces, which provided information about the stages of development of valleys and river-basins; some forms of slopes were studied, e.g. landslide and rockfall, and not the slope as a whole, although the slopes constitute the mojor part of the lands. Karstic forms were studied, both on and under the surface, for their uniqueness and economic importance. Some forms connected with mountain and lowland glaciations were investigated with the help of geologists who determined the stratigraphy and chronology of the Pleistocene. Most often single forms or their groups were investigated, but not all the forms in the given area. In this way, many forms of different origin and age were identified during several decades of years, and on that basis various factors and processes in the modelling of the earth surface were determined. Some particular forms were analysed, set in order; certain types were distinguished and evolutional sequences were established but not all landforms were studied; neither was there any search for the correlationships of these forms, not any detailed systematics was elaborated which would help to arrange these forms into systems and orders.

These investigations resulted, first of all, in a verbal description of particular forms or types of forms. Sometimes these verbal descriptions were supplemented with drawings, profiles, photographs or maps informing about the distribution of these selected forms. They were usually placed against a white background, sometimes against a hypsometric one, but no effort was made to relate them to some other forms of the relief in which they occurred forming a component part.

The geomorphological elaborations of that kind did not and still do not provide a full picture of the relief; consequently, they are of limited value and little utility for practical purposes. In order to get a full picture of the relief it is necessary to study all forms and all elements of the relief, both from the point of view of quality (in what way, under what conditions, and at what time did the particular forms originate) and quantity (dimensions, distribution, frequency, density, rate of origin and development). The investigation of all forms which occur in the given area induces the necessity of marking them on a topographic map, i.e., the application of the method of geomorphological mapping. Geomorphological mapping results in a detailed geomorphological map (1 : 10000—1 : 100000).

THE DEVELOPMENT OF GEOMORPHOLOGICAL MAPPING AND THE PRINCIPLES OF THE PREPARATION OF DETAILED GEOMORPHOLOGICAL MAPS

The first conception of a detailed geomorphological map was presented by S. Passarge in 1914 in the form of a *Morphological Atlas*. It consists of eight maps, 1 : 50000 :

(1) topographic-orographic map with the plant cover,
(2) map of the gradients of slopes (0—5°, 5—10°, 20—35°, >35°),

(3) map of valley forms,
(4) geologo-stratigraphic map,
(5) map of physical resistance,
(6) map of chemical resistance,
(7) petrographic map,
(8) map of the development of the relief.

This is not an only map giving a synthetic knowledge of the relief, but a set of maps informing about the elements of the relief, its development, the structure of the substratum and its susceptibility to weathering processes. The edition of the *Atlas* was preceded by the information about its contents (1912).

At the same time H. Gehne (1912) elaborated a geomorphological map of the environs of Thale. According to him, a geomorphological map should inform about: (a) morphographic features, (b) structure of the substratum, (c) morphology—showing the forms of the relief and their origin. In 1919 J. Smolenski published an article—"on the needs of the Polish science in the field of physical geography"—in which he considered "the study of the relief of the Polish territories and its origin" to be one of the most important tasks and suggested to carry it out by means of "a detailed investigation of the relief of some chosen regions with a different configuration and their elaboration on the basis of detailed cartographic relevées and geologo-petrographic studies". Five years later H. Weber (1924) proposed to prepare special geomorphological maps. The same idea was resumed in 1926 by R. Mayer who prepared a geomorphological map of one of the regions of Istria in the scale of 1 : 25000.

K. Markow (1929) recommended that geological maps should also include morphography, morphogenesis, the age of forms, and the types of the relief. In 1937 Z. Swariczewska proposed a legend to geomorphological maps in the scales of 1 : 50000 and 1 : 500000. In the 30's Smolenski's idea was taken up by B. Zaborski and M. Klimaszewski. The former prepared some maps in the scale of 1 : 100000 (manuscripts), the latter performed a geomorphological survey (1939) of the head of the Biala Woda Valley in the Tatras in the scale of 1 : 25000 (1950). The interest in geomorphological mapping was slight and much more attention was paid to the preparation of general structural-geomorphological maps (Z. Baranski, N. Dik 1934, B. Dobrynin 1935, E. de Martonne 1936, A. Borzow 1938, F. Machatschek 1938).

It was only after World War II that the performance of detailed geomorphological maps on the basis of a systematic mapping of landforms was advanced as a need or even a necessity (H. Boesch 1945, H. Annaheim 1945, 1956, K. Markow 1948, M. Klimaszewski 1948, 1953, 1956, D. Borisewicz 1950, A. Spiridonow 1952, J. Tricart 1954).

During the XVIII Congress of the International Geographical Union in Rio de Janeiro (1956) two conceptions how to prepare a detailed geomorphological map were presented (H. Annaheim, M. Klimaszewski), and geomorphological mapping was considered one of the most important tasks of the Commission on Applied Geomorphology. During the XIX Congress of International Geographical Union in Stockholm eight papers on mapping were already submitted. It also was established that the schemes of geomorphological maps prepared in Switzerland, USSR, Poland, France, Czechoslovakia, Japan, Belgium and Hungary independently one on the other differed in their content and the manner of its representation on the map. In connection with this, a Subcommission for Geomorphological Mapping was created (with M. Klimaszewski as Chairman) and its tasks were as follows:

(1) to introduce in geomorphology the method of geomorphological mapping, the result of which is a detailed geomorphological map,

(2) to adopt a uniform system and principles of the preparation of detailed geomorphological maps to ensure their comparability,

(3) to provide the national economy, by means of a geomorphological map, with a detailed picture of the relief, which is an important element of the geographical environment, and thus to facilitate a rational utilization of the earth surface. The geomorphological map should be used in economic planning, both local and regional, together with geological, hydrographic, hydrogeological, mesoclimatic, palaeological and topographic maps.

Many papers were written (cf. the Literature) and several sessions of the Subcommission of Geomorphological Mapping were organized in Poland (1962), France (1962, 1963), Great Britain (1964), Czechoslovakia (1965), Belgium (1966), and the USSR (1967) to realize these tasks.

During these sessions:

(1) various geomorphological maps were evaluated, and a clear distinction was drawn between detailed geomorphological maps and the morphographic, morphographic-morphometric, morphogenetic maps and those of relief types as ones which do not provide a full picture of the relief and its development, although the latter maps are often erroneously called morphological maps,

(2) discussions were devoted to and decisions were made on the conception and principles of construction of detailed geomorphological maps to ensure their comparability independent from the author and the country, special attention being given to:

(a) the determination of the scale (1 : 10000—1 : 100000) and the scope of content of detailed geomorphological maps,

(b) the conception of a uniform legend of a detailed geomorphological map or the principle of its elaboration,

(c) the principles of a uniform construction of a detailed geomorphological map to make it readable and comparable.

(d) estimation of the utility of detailed geomorphological maps for practical purposes.

During the session of the Subcommission in Cracow (1962) the participants, representing fifteen countries, agreed upon the following principles of preparing detailed geomorphological maps:

1. A detailed geomorphological map must be a result of geomorphological mapping in field; it is recommended to use aerial photographs-photointerpretation.

2. Detailed geomorphological maps are elaborated in the scales ranging from 1 : 10000 to 1 : 100000, because in such scales the relief and its peculiarities can properly be represented.

3. A detailed geomorphological map should present a full picture of the relief enabling the user to learn the character of this relief, to reconstruct its past and to foresee further tendencies of development. The map should inform about the distribution and correlations of forms of a certain appearance,

size, origin, and age; it must include morphographic, morphometric, morphogenetic and morphochronological data.

4. All the forms investigated should be marked on the map by means of coloured signatures true-to-area. By means of signatures and colours information is imparted on the outlook, size, origin and age of the landforms.

5. The determination of the age of the forms is necessary because it introduces the chronological order in the contents of the map and helps to reconstruct the geomorphological development and to predict further developmental tendencies.

6. The lithological data should be marked with special signature;, preferably on the background of the map.

7. The legend should be arranged in a genetic-chronological order.

8. The detailed geomorphological map is of great importance for the development of geomorphology and for the investigations in the field of regional geomorphology carried out in territories differing in structure and climatic conditions. It is also very important for practical application. On the basis of these maps, special maps may be prepared e.g. those enabling the estimation of the quality of the site or representing the distribution of forms and their groups favourable or unfavourable for the particular branches of national economy. Therefore, a detailed geomorphological map is both of scientific and practical value. This map may also be helpful in preparing pedological or engineering maps, plans of urban and industrial development, plans of agricultural ameliorations, water management, etc. These principles, determined in 1962 and developed during subsequent sessions of the Subcommission of Geomorphological Mapping, facilitated the elimination out of the detailed geomorphological maps the morphographic, morphographic-morphometric and purely morphogenetic maps (without dating of landforms). They also facilitated the understanding between representatives of two different trends in the construction of a detailed geomorphological map. French, Czechoslovakian and Hungarian geomorphologists propagated the construction of structural-geomorphological maps, continuing the idea of E. de Martonne (*A geomorphological map of France 1 : 1000000*). They considered the lithological-structural unit to be the basic element, and on that background

represented with coloured patches or different colours they examined the landforms of defined sizes, origin and age marked by means of coloured signatures. These maps inform about the relation between the landforms and the structure of the substratum, but they prove useless at the reconstruction of the development of the relief.

Polish, Russian, Rumanian and German geomorphologists consider the form to be the basic element; therefore it is not the structure but the relief of definite dimensions, origin and age of the landforms, which is the element of prime importance in their maps. Such maps give the right information about the character of the relief and enable the reconstruction of the geomorphological development; on the other hand, they fail to inform about the relation between the relief and the geological structure. However, to this purpose it is enough to compare the geomorphological and the geological map. A closer cooperation of the representatives of these two trends led to the limitation of the geological content in geomorphological maps, and the consideration of the structural-lithological data in cases of structural, tectonical and accumulative landforms. They also agreed upon the scope of the contents, the principles of mapping and elaboration of detailed geomorphological maps. Owing to this, detailed geomorphological maps are more and more comparable as far as the scope of the contents is concerned. However, the legend of these maps and the signs introduced in the detailed geomorphological map representing the located forms differ considerably; in this respect the geomorphological maps are incomparable.

All geomorphologists agree that detailed geomorphological maps should be made comparable, and that a uniform legend should be elaborated. This is a very ambitious idea but difficult to be put in practice, because the authors seem to be attached to the signatures of their invention, especially with reference to small regions. To elaborate a uniform legend it was necessary to think over colours and signatures for very numerous forms of different dimensions, origin and age occurring all over the world. A special working group was appointed, headed by M. Klimaszewski, to work on the legend; its members were as follows : N. Boszenina, J. Gellert, F. Joly, E. Scholz, and S. Gilewska (secretary). During several sessions that team worked out a uniform legend for the particular geomorphological maps of the world (1 : 25000—1 : 50000). It was submitted to the IGU Congress in New Delhi (1968), and published in Folia Geographica, series geographicaphysica, Vol. 2, Cracow 1968. The legend includes explanations in 5 languages (English, Russian, French, German and Polish) of 500 signatures of landforms represented on detailed geomorphological maps.

In this way the Subcommission of Geomorphological Mapping realized the major part of the programme presented during IGU Congress in Stockholm (1960). Indeed, the activities of the Subcommission should be brought to an end, and some new problems and tasks ought to be proposed, to inspire the geomorphologists to action in a new direction. However, in the IGU there is a tendency to change the Commission into a possibly permanent institution, the task of which is to carry out studies rather than to initiate them. The cause lies in the fairly poor ingeniousness of the geographers, geomorphologists included, lack of new problems and ideas, which would demand cooperation of scientists from different parts of the world. This is an alarming phenomenon.

Accordingly, the Subcommission on Geomorphological Mapping was not dismissed by the IGU Congress in Delhi (1968) but changed into the Commission on Geomorphological Mapping. That Commission took as its main task the elaboration of the "Manual of detailed geomorphological mapping" and of the "Legend to the International Geomorphological Map of Europe" in the scale 1 : 2500000. That manual consists of the papers written by 20 geomorphologists. They submit information on mapping and geomorphological maps, as well as on morphographic, morphometric, geological and other maps concerning different elements of the geographical environment. They also impart knowledge on the methods of laboratory investigations of sediments and soils on geomorphological regionalization, and on the utilization of geomorphological maps for practical purposes, i.e., information useful for every geomorphologist. On the other hand the chapter on "Geomorphological mapping in areas of various structures and climates" contains information typical of manuals. The scope of that knowledge is fairly narrow and squeezed on 60 pages, while the role of the structure has been altogether lost. The "Manual" is a sort of a compendium of general geomorphology rather uneven, though on the whole useful, but it did not require the supervision of the Commission and consideration as one of its main tasks.

The other task, i.e. "the elaboration of the legend to the Geomorphological Map of Europe in the scale of 1 : 2500000" has nothing to do with geomorphological mapping.

The preparation of a legend to a general map is a purely editorial work and consists in compilation, and not in field investigation, i.e., mapping.

THE COURSE OF THE ELABORATION OF A DETAILED GEOMORPHOLOGICAL MAP

(It is necessary to know the essentials of geomorphology and the key to the sign of a detailed geomorphological map).

CHOICE OF TERRITORY

The choice of territory may be dictated by scientific or by practical reasons: it may cover the whole area of a given country, with a division into sheets (there were such plans in Poland, Belgium, the Netherlands, and other countries).

PREPARATION

1. Preparation of topographic groundworks

(a) contour maps for field investigations should be cut into sheets 14×20 cm and stuck on cardboards.

(b) contour maps for fair copy.

2. Preparation of aerial photographs of the chosen territory and of a stereoscope.

3. Preparation of geological maps of the chosen territory.

4. Preparation of a list of the geomorphological, geological and geographical literature concerning the chosen territory, and its wider region.

5. Preparation of the equipment for field investigations : Bag—geological compass, altimeter, clasimeter, clinometer, hammer, spade, knife, magnifying glass, metal and sackcloth

tapes, cailleaux disc, slide calliper, muriatic acid. Notebook 14.5 by 21.0 cm, pencils, coloured pencils, rubber, camera, plastic bags, metal cylinders, glass pipes to take samples of soil and fossil plants.

ELABORATION OF THE MATERIAL PREPARED

1. Detailed analysis of topographic maps (main features of the relief, hypsometric relations, degree of fragmentation, recognition of forms of different origin and order).

2. A detailed analysis of aerial photographs under the stereoscope (a faithful and plastic picture of the relief helping to recognize many geomorphic details and especially the range of many landforms and their elements).

3. A detailed analysis of geological maps and profiles (a possibility to recognize and define the relation between the landforms and the geological, i.e., lithological and tectonical structure).

4. Acquaintance with the geomorphological and geological literature concerning the territory investigated; preparation of notes and abstracts as well as of a list of the literature.

5. Marking on a topographic map the data concerning the distribution and dimensions of the landforms identified during the analysis of topographic maps and aerial photographs, as well as the data concerning lithology (quality and resistance of rocks) and tectonics (faults, joints).

6. Preliminary plan of routes and research points should be prepared on the basis of literature, the results of an analysis of aerial photographs, topographic maps and preliminary exploration in field.

FIELD INVESTIGATIONS—MAPPING

1. A detailed recording, location and classification of all landforms with respect to their morphographic and morphometric features, origin (on the basis of an analysis of the forms and deposits as well as of their relation to the geological structure), and age (on the basis of an analysis of the forms and deposits in relation to the dated rocky and depositional

formations)—cf. the chapter on the method of geomorphological investigations.

2. Mapping of the landforms, which consists in marking on a topographic map all forms found during field investigations in the given territory by means of accepted signatures true-to-surface. These forms should be described, measured and classified both genetically and chronologically. Mapping requires the ability to read a topographic map and locate precisely each spot and locality.

3. Depending on local conditions, the following methods may be used in mapping :

(a) the profile method—of geomorphological cross-sections—and taking routes along previously determined more or less parallel lines, and performance of detailed studies in some determined places or sections;

(b) the skeleton method—taking routes along valleys and divides between them—mapping of landforms at the bottom of the valley—on the slopes—on the top surface; the order is twofold : first the bottom, next the slopes, and then the top surface, or rather in sections ; section I: bottom—slope—top surface; section II: bottom—slope—top surface, etc.;

(c) the method of studying the limits of the range of landforms; taking routes along the limits (edges, zones) separating the landforms of different origin and age (e.g. along the ridges of river terraces, cliffs, landslide niches, etc.). Photointerpretation may replace the tracing of the range of landforms;

(d) the point method of recording, locating and describing the landforms from suitable look-out points.

The skeleton method and that of studying the limits are used in mountainous regions, while the point method is used in inaccessible regions. The profile method and that of studying the limits are used in the lowlands.

4. The conditions of mapping are multiple:

(a) easy—in open areas,

(b) difficult—in partly forested territories,

(c) very difficult—in territories fully overgrown with forests.

The rate of mapping depends on:

(a) the character of the relief: plains; uplands; low medium, and high mountains; distinctness of forms; number and distribution of natural outcrops and man-made exposures informing about the structure and origin of landforms,

(b) climatic conditions (frequency of precipitation, fogs, intensity of frost),

(c) biological conditions (character of plant cover, marshlands, plagues of mosquitoes, flies, etc.).

5. Simultaneously with the mapping of landforms, descriptions and data on their structure—the contents—are recorded in the field notebook, or on the tape. The text is complemented by profiles of the landforms, cross-sections of outcrops, perspective drawings and photographs.

6. Geomorphological mapping does not consist in mechanical recording and location of the forms identified; it should be accompanied by conceptual thinking, reconstruction of entire forms on the basis of fragments, reconstruction of the participation of various factors and processes in the modelling of the relief on the basis of the knowledge of the forms already investigated, reconstruction of the stages of development of the relief of the area investigated, and the estimation of the utility of the landforms for various branches of economy.

ELABORATION OF A DETAILED GEOMORPHOLOGICAL MAP IN STUDY ROOM

1. Performance or ordering the performance of mineralogical petrographic, granulometric, morphometric, morphoscopic, palynological and other laboratory investigations of the samples collected in field.

2. Elaboration of the fair copy of a detailed geomorphological map, i.e., transmission of the geomorphological contents from the field maps on the map destined to be a fair copy (use being made of the descriptive data recorded in the notebook, data drawn from the literature, and the results of laboratory investigations)—by means of multicoloured signatures true-to-scale. The map should give a full and measurable picture of the investigated territory conceived morphogenetically and chronologically, i.e., developmentally. It must give an idea on the distribution of the landforms of a determined origin and age, and impart information about their size and

morphometric characters. The map must be constructed in such a way that the successive stages of the geomorphological development of that area and its further development tendencies could be read from it.

3. Work on the fair copies of the geologic-geomorphological profiles characteristic of the area investigated.

4. Writing an explanation to the detailed geomorphological map: Area—status of studies—method—general orographic and geological characterization—description of landforms, which may be executed in the morphogenetic order (according to factor and processes), or in the chronological order (from oldest to youngest), or in the topographic order (along routes)—opinion on the development of the relief and further developmental tendencies—estimation from the economic point of view, literature, figures.

5. The elaboration of detailed geomorphological maps on the basis of mapping in field is a work taking much time and labour, and requires a greater creative and even physical exertion than do the investigations in dynamic geomorphology. That is why studies in the field of geomorphology and dynamic geology recently arise more interest among geologists, sedimentologists, hydrologists and soil scientists who also undertake them, while geomorphological mapping is exclusively the domain of geomorphologists.

PRINCIPLES OF THE CONSTRUCTION OF A DETAILED GEOMORPHOLOGICAL MAP ELABORATED BY M. KLIMASZEWSKI (1956)

A detailed geomorphological map is the result of geomorphological mapping. All landforms of a determined provenance and age are marked on a hypsometric background by means of coloured patches and signatures. The colours inform about the factors (stream water, glacier, wind, gravity etc.) which in some determined time (geological age modelled the given area due to a destructive or constructive activity (process). For example, the forms which originated due to a destructive activity of rivers and of denudative processes in the Holocene are marked with red signatures, and the forms created in result of the constructive activity of rivers and denudative processes in the Holocene are marked with blue signatures.

The landforms created in consequence of a destructive activity of rivers and denudative processes in the Pleistocene are marked with orange signatures, while those formed due to the constructive activity of rivers and denudation processes in various periods of the Pleistocene are marked with signatures coloured green in its various shades.

Grey signatures denote the forms created due to the destructive activity of rivers and denudation processes in the Neogene, and the respective strata are determined by means of letters. For the principles of the application of colours cf. Geomorphological map of Poland prepared by Physical Geography Department, Institute of Geography and Spatial Organization, Polish Academy of Sciences, Cracow, M. Klimaszewski, *Geomorfologia*.

Thus, on geomorphological maps the colours inform about the destructive or constructive activity of various factors in different geological periods, and the signatures about the results of that activity, i.e., about real forms. The colour informs simultaneously about the provenance and age of the particular forms, and that principle characterizes the Polish version of the geomorphological map. The determination by means of colours of the provenance only, or of the age only, entails disadvantages for the clearness of the map and produces a map in which either the origin or the age are most strongly emphasized, and not these two characters together.

A detailed geomorphological map also informs about the morphometric characters. The signatures, by means of which landforms or their elements (e.g. slopes) are represented, are true-to-scale and therefore inform about the real dimensions of the landforms. The absolute and relative height may be read from the contour lines because the map is executed on a hypsometric background. The depth and height of small landforms (e.g. gorges ravines, ridges, terraces) is marked by means of lines of variegated thickness, size of dentation, etc. The gradient of slopes is marked by means of the shades of the colour denoting the origin and age of the form of which the slope is an element (slope of valley, threshold, moraine ridge, etc.)

In this way, the Polish version of a geomorphological map informs about the outlook and size (morphometric characters) of landforms by means of signatures and about the origin and age of forms by means of colours; thus it contains all data necessary for the knowledge of the character of the relief and its development. That development should be considered (reconstructed and foreseen) on the background of geological structure and in connection with other elements of the geographical environment changing during their development (climate, water, soil, plant cover).

The list of forms and signs applied in this version of the detailed geomorphological map was published in the *Problems of geomorphological mapping*, Geographic Works of the Geographical Institute of the Polish Academy of Sciences, No. 46 (1963). That list was reprinted in *Encyclopedia of Geomorphology*, published in 1968, edited by R. Fairbridge with D. St. Onge's paper *Geomorphic Maps*.

The elaboration of the geomorphological map of Poland was recognized by the First Congress of the Polish Science (1950) as one of the main aims of geography. The mapping, in which geomorphologists of all geographical centres participated, was financed up to the year 1954 by the Fund for the Promotion of Scientific and Artistic Creation, and in the period 1954—1967 was included in the research programme of the Geographical Institute of the Polish Academy of Sciences in Poland. In the course of these years about 1/5 of the territory of Poland was mapped, and an efficient cooperation was established among geomorphologists. The series of 67 sheets of the geomorphological map scaled 1 : 25000 and 1 : 50000 published up to 1968 includes the paper by M. Klimaszewski *Problems of the detailed geomorphological map*.

PRINCIPLES OF THE CONSTRUCTION OF A *DETAILED GEOMORPHOLOGICAL MAP* ELABORATED BY THE TEAM COMPOSED OF N. BASHENINA, J. GELLERT, F. JOLY, M. KLIMASZEWSKI (CHAIRMAN), E. SCHOLZ, AND S. GILEWSKA (SECRETARY)

The members of the team appointed from among the participants of a conference of the Subcommission of Geomorphological Mapping of the International Geographical Union stated that a detailed geomorphological map should be complete (i.e. informing about the morphographic, morphometric, morphogenetic, and morphochronological data), plastic (because the relief is plastic), readable (signature should be simple and reminding on the forms they represent), true-to-scale (the size of forms represented precisely in the given scale by means of true-to-scale signatures, and small forms by means of symbols), and useful (enabling the location and spatial analysis of forms of various economic utility).

A map constructed according to these principles should give a possibly plastic and clear picture enabling the determination of the character of the relief and the reconstruction of its development as well as a prognosis of further developmental tendencies and the possibility of the estimation of its economical utilization.

In reference to the principles mentioned above it was stated that the geomorphological content should be represented on that map by means of coloured patches and coloured signatures. Considering the high cost of printing of multi-coloured maps (the Polish legend proposes 13 colours and their numerous shades, the French one 10 colours, and the Russian one 15 colours and very many shades) it was decided to introduce some restrictions in the number of colours. The following colours are proposed for the designation of genetic sets of forms:

red —for endogenetical (neotectonic and volcanic) forms and those conditioned by structure,

brown —for the destructional forms of denudative, fluvial, fluvio-denudative, karstic and suffosional origin,

green —for the accumulative of denudative, fluvial, fluvio-denudative, karstic and suffosional origin,

violet —for the destructional forms of glacial, nival, criogenetic and thermo-karstic origin,

violet+brown—for the destructional forms of glacio-fluvial origin,

pink —for accumulation forms of glacial, nival and criogenetic origin,

pink+green —for accumulation forms of glacio-fluvial origin,
blue —for the forms of limnic and marine (talasogenetic) origin,
yellow —for the forms of aeolian origin,
black —for the forms of organogenetic and anthropogenetic origin,
light blue —for waters and glaciers.

Thus, the colour informs about the origin of landforms, while the shades of colours are to inform about their age. Dark shades were proposed for younger forms, light shades for older ones. The age of forms may also be marked by letters used in the international geological nomenclature. Thus, by means of colours and their shades the map will inform about the origin and age of landforms. That map will be less readable than the geomorphological map in the Polish version, but being cheaper it will be easier in realization.

As to the signatures representing in a true-to-scale way the landforms differing in outlook, dimensions and origin it was decided to make them simple, readable and resembling the landform represented. There should not be introduced any complicated signatures bearing the character of geometrical or plant ornaments.

The authors tried to preserve point signatures for accumulation landforms, and linear signatures for destructional landforms.

The morphometric data were introduced in the background of the map as contour lines (grey lines) and grey screens of various density denoting different classes of the gradient of slopes, as well as various signs (different thickness of lines, size of dentation, etc.) which complement the coloured signatures. To some extent, the map also informs about the lithological data on the substratum by means of additional signs introduced in the contents of the signatures representing the structural and accumulation landforms.

A map thus constructed gives a picture of the relief in both genetical and chronological aspect allowing the determination of the character of the relief and the reconstruction of its geomorphological development; it also enables an evaluation of the relief from the point of view of various branches of economy. A list of landforms and signs is included in the *Project of the unified key to the detailed geomorphological map of the world* published in Folia Geographica, Series geographicaphysica, Vol. II, Krakow 1968.

THE SCIENTIFIC VALUE OF A DETAILED GEOMORPHOLOGICAL MAP

A detailed geomorphological map gives precise picture of the relief in its dynamic aspect, and for that reason it enables the reconstruction of the development of the relief, the reconstruction of the developmental stages, the cognition and evaluation of the participation of various factors and processes in its origin and transformation. A map representing the distribution of all landforms of determined dimensions, origin and age facilitates the search for their connections, enables the regionalization i.e. distinction of morphogenetic regions, and defines their exact character. It also facilitates a proper development of comparative studies. It allows to compare the landforms already created, and those originating and developing in areas of a similar structure (e.g. in medium-sized granite limestone or sandstone mountains), but different climatic conditions, and in areas of a different structure but similar climatic conditions and during a similar span of time of their formation. The performance of a geomorphological survey in areas of different structure and climatic conditions facilitates a far more exact cognition of the development of the relief, and of the principles governing the development. Geomorphological maps of various areas allow to compare the equal-aged landforms, e.g., Holocene ones, with respect to both quality and quantity (dimensions, density of distribution, rate of development). Such a procedure will help to become acquainted more precisely with the role of the climate in the shaping of earth's surface, and to distinguish the climatic types of the relief, i.e., the morphoclimatic zones and regions. This way is more objective and more efficient than the comparison of topographic maps. A geomorphological map is doubtlessly of great importance for the development of geomorphology; the lack of such a map has hitherto checked that development in a considerable degree.

PRACTICAL VALUE OF A DETAILED GEOMORPHOLOGICAL MAP

All sections of the economy are linked up—in a greater or lesser degree—with the geographical environment including the configuration of the earth surface.

A detailed geomorphological map providing an exact and measurable picture of the relief meets the requirements advanced by various fields of economy tending to a more rational utilization of landforms. The configuration of the earth surface is of greatest interest to agriculture, settlement, communication, hydrologic engineering, tourism, recreation, and the management of resources.

Agriculture develops best in the territories which are flat, not much inclined, and slightly dismembered or transformed by contemporary morphogenetic processes. Mechanical cultivation of potatoes is possible in flat territories, inclined up to 4°; for the cultivation of beetroots a gradient of 7° is extreme, and for cereals 15°; it is recommended that on slopes inclined at 7°—11° the fields should form bands, at 20° ploughed terraces, and at higher gradients narrow terraces should be formed.

A detailed geomorphological map informs about the distribution of plains and slopes of different gradient, favourable, or unfavourable for, or even impeding agriculture, e.g., rain-washed and depositional landforms (soil erosion); landslides; soil-creeps and suffosional, aeolian and flooded forms; those developing at present and testifying to the occurrence of unfavourable relief-forming processes in precisely located places and requiring melioration.

The knowledge of the distribution of territories favouring or disfavouring agriculture marked on a detailed geomorphological map, and the knowledge of agricultural utilization marked on a map of land utilization enable the estimation of the rightness of the contemporary utilization and the planning of a more rational management. It is accepted that plains and slopes inclined up to 11°may be cultivated as arable fields, the slopes inclined at 11°—20° should be cultivated for fodder or taken under orchards and woodlands, and the slopes inclined more than 20° should be occupied by forests.

The building and extension of towns, settlements and industrial enterprises require flat or slightly inclined (up to 6°) territories. More inclined slopes cause difficulties; if these are to be mastered, one should take into account whether these investments pay; one can build on slopes, even on very steep ones, by cutting terraces and constructing rests, but the costs are very high.

The building plans should take into serious consideration the distribution of territories threatened by landslides, soilcreeping, stone-falling, avalanches, and of rain-washed, suffosional, karstic and flooded areas. When the morphogenetic processes recorded on a geomorphological map were ignored there resulted great damages and losses.

The impact of the relief on the development of lines of communication is obvious, and has been well known since long. In mountainous territories the roads run along valley bottoms, climb the slopes in serpentines, or are led through the ridges in long tunnels. In the uplands, valleys and levelled top surfaces are utilized; in the lowlands the roads run on drier bulges.

The best conditions for road building are found in flat, and slightly inclined and dismembered territories. Railway lines can conquer slopes inclined up to 2°, highways up to 4°, arterial roads up to 5°, other roads for motor and horse-drawn vehicles up to 7°, roads for tractors up to 15° (K. Sokalski). A geomorphological map informs about the distribution, extension and segmentation of plains and slopes of various categories. It also informs about the distribution of rocky walls subject to weathering and falling (which entails the need for building tunnels, galleries, trenches or setting networks); about the distribution, dimensions and activity of landslides and slopes threatening with slides when undercut by roads; about the distribution of soil creeps and rain-washed slopes, ravines, debris cones, avalanche grooves, suffosional and karstic areas, deflation and aeolian depositions of sand, dust and snow, and about frost-logged or freezing areas.

The roads in the mountains are their "Scars" visible from afar and impairing the beauty of the natural landscape. However, they are a necessity. To plan and build roads skilfully and in harmony with the environment it is necessary to acquire a thorough knowledge of the relief.

Hydrologic engineering takes advantage of the conditions created by the configuration of the earth surface. To build water dams it is necessary to have a thorough knowledge of the configuration, especially of the valleys, and of the contemporary morphological processes within the precincts of the slopes of the reservoir, and of the whole catchment area.

Water dams are usually located in the narrow stretches of river gorges—except antecedent ones, which are often subject to young tectonic movements, and the gorges cut in karstificating limestone rocks. The knowledge of the slopes bordering a retention reservoir, and of the landforms occurring in the vicinity, enables the prediction of a revival of morphogenetic processes unfavourable for the stability and permanence of the reservoir, i.e. landslides, rockfalls, soil-creeps, rain-washing, the revival of which brings about catastrophic results.

The undercutting of slopes and slope covers by water waves in the reservoir causes the formation of several-levelled abrasion terraces, and weakens the stability of slopes. All these processes accelerate the filling up of the reservoir. This happens mostly due to the rivers and streams which carry their waters into the retention reservoir. A geomorphological map informs about the vitality of the streams flowing in the catchment area, and distinguishes the valleys which are cut and the bottoms which become levelled.

Flood control and river training operations must also be preceded by the acquisition of a thorough knowledge of the relief, especially as regards the bottom of the valley and of the river bed.

A geomorphological map informs about the character of the bed (rocky, alluvial), its course (straight, winding, meandering, stress-like); about old river beds; about the sills occurring in them; about river banks undercut or raised; about the range of flooded territories covered with water and silt at different stages of high water (cf. Japanese geomorphological maps by M. Oya.)

Development of sea-shores for economic and recreational purposes requires a thorough knowledge of their configuration and of the processes to which the coast is subject. A geomorphological map informs about the distribution and dimensions of the sections with the cliff both living and dead, as well as of the distribution and area of beaches, and the destructive and building tendencies.

While setting aside touristic and recreational territories it is the configuration of the substratum which is taken into consideration above all. The touristic and recreational movement is concentrated mostly in the areas of a variegated relief (mountains, lakelands, uplands) and on the sea-coast. To the most attractive territories there belong the mountains with a glacial and karstic relief, monadnocks, deep ravines, canyons and moraine hills. Most of these territories are not suitable for agriculture. Their accessibility, freedom of environmental pollution, and safety from vehement morphogenetic processes (e.g., floods, rockfalls, and avalanches) is of great importance.

A geomorphologic map also informs about the distribution of landforms built of certain deposits which are important as raw material in building, e.g., esker ridges are built of sand and gravel; kame hummocks and tableland mainly of sand and sometimes of organic muds; frontal moraine ridges often of rocky blocks; stagnant basins are filled with varve clays; sandre, terrace and dunal plains are sources of sand; river cones and terraces in the mountains are built of gravel and sand.

The knowledge of the relief and of the landforms favouring or disfavouring various branches of economy and of their distribution enables a better planning and a proper management of the territory investigated, while the knowledge of the laws governing the development of young forms and of the conditions of that development makes possible the control of the processes and forms unfavourable for economy (e.g., landslides, gorges, suffosional and karstic subsidences, etc.); it facilitates the work in the field of mastering and rational utilization of the geographical environment.

A detailed geomorphological map may serve as a basis for the construction of derivative maps. These may represent the distribution of certain forms only, e.g., those disfavouring a given branch of economy (landslides, gorges, debris cones, stream beds deepened by erosion). Other derivative maps may show the quality class of the particular territories.

The principles of constructing geomorphological site-qualifying maps are as follows : All the landforms established in the area investigated and recorded on a geomorphological map are evaluated from the point of view of their usefulness of the given branch of economy, e.g., the slopes of various inclination, alluvial plains, landslides, gorges, suffosional and karst funnels—for agriculture, building industry, communication, and others. In this way every landform is qualified as favourable (S) for the development of the given branch of economy, unfavourable (N) for it, or favourable on condition of certain melioration measures being applied (W). Next, by using only three signatures, one denoting the favourable forms, the second conditionally favourable ones, and the third unfavourable forms, the outlines of landforms and elements of forms of different qualification are transferred from the geomorphological map on the geomorphological-site-qualifying map. The latter informs about the distribution, dimensions, and compactness of the areas of different usefulness for the given branch of economy from the point of view of geomorphology.

On geomorphological-site-qualifying maps there is marked the distribution of favourable and conditionally favourable areas only.

The utilization of a detailed geomorphological map by various economic and planning institutions (e.g. in Poland, the USSR and France) proves that it is not only of scientific and theoretical but also of practical value. This double importance of a geomorphological map both for science and for practical purposes obliges the geomorphologist, who investigates and maps a certain area, to look at every landform, process and phenomenon from two sides : the theoretical and scientific point of view (when, how and under what conditions did the given form originate), and from the practical, economic point of view (what value does it represent for various branches of economy and how can it be utilized). This double look and double aspect of investigations favours the development of geomorphology.

THE HIGH TATRAS*

The part of the detailed geomorphological map of the Polish Tatra Mountains includes a great part of the Rybi Potok valley, the Pieciu Stawow Polskich valley with the part of the Roztoka valley and the Kozia valley, being the part of the Sucha Woda valley. The area mentioned above is composed of granodiorites, dissected by the dense system of joint cracks and faults, with the directions NW-SE and NE-SW. The map presents a high-mountain relief, with the strong transformation of Pliocene fluvial and denudational relief by the Pleistocene glaciations. During the three periods of the glaciation (Mindel, Riss, Würm) the glaciers modified the relief of the upper parts of this valley, which were not rejuvenated during the upper Pliocene and lower Pleistocene. These uppermost, non-rejuvenated parts of the valleys were transformed into glacial cirques, and the rejuvenated parts of valleys were transformed into glacial troughs. Above the upper sides of the glacial troughs are preserved few fragments of the level, corresponding with the non-rejuvenated parts of the valleys. This transformation was based on the widening and overdeepening of the preglacial river valleys, coinciding with the NE-SW directions of the fissures. The great glacial forms, overlying the fluvialforms, are represented by the glacial cirques and troughs, Some types of the glacial cirques were distinguished, depending on the degree of glacial transformation:

(a) the concave-bottom cirque, with the overdeepened bottom occupied by the lake waters (the cirques of the Czarny Staw above the Morskie Oko and of the Zadni Staw):

(b) the flat-bottom cirque, with some levels at the bottom, and with the rôches moutonnées fields (at the Za Mnichem valley) are covered with morainal and talus deposits (Pusta, Buczynowa, Swistówka and Kozia valleys).

(c) the slope-bottom cirque, coated with debris (under Mieguszowiecki, under Cubryna, under Miedziane and under the Szpiglasowa).

* Please refer if necessary to volume XII Folia Geographica series geographica-physica for the relevant map of the Polish Tatra Mountains.

The bottoms of the cirques are bordered by the rocky walls at three sides. The walls are dissected by numerous ravines with late-glacial talus cones at the outlet. Toward the valley-through the basins of the cirques are closed by the smoothed rocky bars with the rôches moutonnèes dissected by rapid streams during the postglacial period. The flat-bottom and the slope-bottom cirques are bordered from the valley by the ridges of recession and moraine.

The Pieciu Stawów Polskich valley is a glacial cirque. It originated as a result of transformation of a whole system of the non-rejuvenated preglacial valleys. During the glacial periods, these valleys were situated above the perennial snowline and became the reservoirs of firn- and ice-masses. The lateral valleys were transformed into cirques and the main valley, with the length of five kilometres, was widened and overdeepened: below the rocky steps (the Zadni Staw, the Czarny Staw) and at the places, where ice-masses were jointing and their thickness was increasing (the Dlugi Staw, the Przedni Staw). The bottom of the Pieciu Stawów valley is either of the rôches moutonnées, or is covered with morainic material. The bottom of this valley slopes rapidly as steep, abruptly cross wall of the Siklawa, of the height of 250 m, to the Roztoka valley. This valley was rejuvenated during the upper Pliocene and lower Pleistocene. The height of the outlets of the hanging Buczynowa and Swistówka valleys is the same, about 1700 m a.s.l.

The glacial troughs of the Rybi Potok valley and of the Roztoka valley result from an incomplete transformation of the previous V-shaped river valleys. The bottoms are widened and the lower parts of the slopes up to the height of 200—250 m were undercut and became steeper due to the erosional activity of glaciertongues supplied with the firn reservoirs of a differentiated capacity. The upper parts of the slopes, until the ridges, are of a little lower inclination. The slopes are dissected by a dense system of the ravines, coinciding with the direction of the fissures. At the outlet of the ravines, the talus cones and heaps were accumulated, following the removing of ice-cover from trough-sides. Such cones are of different age: the highest the cone inside the valley, the latest the period of its accumulation and consolidation. The cones are covering the lower parts of troughs sides and the lateral moraine deposits.

The only places, where the lateral moraine ridges are preserved, are: the small-transformed Za Mnichem valley, the level under the Mnich Peak and the eastern slopes of the Opalone ridge, at the height of about 200 m above the Rybi Potok valley. The bottoms of both troughs (the Roztoka valley and the Rybi Potok valley) are of some different configuration. At the narrow trough of the Roztoka valley the talus cones reach the talweg, and the stream hardly forces through the talus of debris and stone-falls. At some places the profile of the valley returns to V-shape one. At the wide trough of the Rybi Potok valley a great part of the valley-bottom is occupied by a plain of the ground moraine. Its surface is diversified with numerous dead-ice hollows, resulting from the areal deglaciation, before the stadial re-advancing of the glacier down to the Szalasiska. The deglaciation—the deterioration of the glaciers during the last glaciation (30—10 thousand years ago) proceeded not uniformly, the periods of regression of the glaciers were interrupted either by the periods of stagnation, or, even of the re-advancing of terminal face of the glaciers. This is evidenced by the recessional end moraines, preserved at the wide bottom of the Rybi Potok valley: above the Szalasiska (1380 m a.s.l.), at the rocky bar of the Morskie Oko lake (1406 m a.s.l.). When the compact glacier tongue had broken up into individual ice-lobes, the four cirque glaciers were existing: at the Czarny Staw (1590 m a.s.l.), at the Za Mnichem valley (1800 m a.s.l.), at the small depression under the Cubryna (2080 m a.s.l.) and at the Mieguszowiecki cirque (but the moraine did not preserve there). The ridge of end-moraine above the Szalasiska (1380 m a.s.l.), formed during the re-advancing of the glacier, grades into fan-or-terrace, coating the bottom of the gully, cut into ground moraine deposits by the from-glacier waters. The fan-or-terrace level, grading into low terrace below the Szalasiska moraine, is also connected with the morainic ridges of the Morskie Oko. The next phases of glacier recession are not marked by the new terraces, because the Morskie Oko lake smoothed the variability of the discharges of proglacial waters. The Morskie Oko lake is not a cirque lake. It occupies a deep depression, eroded at the upstream part of the glacial trough by the ice-masses, moving down from the cirques: of the Czarny

Staw, under theMieguszowiecki Peak and from the Za Mnichem valley. The Morskie Oko is a result of confluent overdeepening.

At the Roztoka valley the only small remnant of ground moraine is preserved. It is hanging about 20 m above the bottom of the valley, modelled by the stream. The ridges of end moraines are not evidenced at the Roztoka valley, because of its hard transformation by talus cones. In contrary, the morainic forms are well-preserved at the Pieciu Stawów Polskich valley. The outlet of this valley is closed by the huge ridge of end moraine (1690 m a.s.l.) with the numerous dead-ice hollows. This ridge of moraine covers the rôches moutonnées. Synchronic with this ridge are the external end moraine ridges at the outlet of the Buczynowa valley (1700 m a.s.l.) and the Swistôwka valley (1705 m a.s.l.). The end moraines, closing the slope-bottom cirques under the Miedziane (1730 and 1790 m a.s.l.), the ridge preserved above the step of the Pod Kolcm valley (1850 m a.s.l.) and at the Pusta valley (1970 m a.s.l.) and the ridges of talus moraine at the cirque under the Szpiglasowa Pass (1960 m a.s.l.), inform about further, stadial deteriorate of the glaciers. The recession of the glaciers followed the exposing of either smoothed, mammilated surface with rôches moutonnèes or morainic surface, and dissecting the substrata by the proglacial waters. At the Pusta and Pod Kolem valleys, the melting of ice-masses caused an irregular accumulation of the huge talus and debris ablation moraine.

At the Holocene, the processes of the weathering, falling down, and over-accumulation of talus cones are still continued. The intensity of this processes depends on the orographical and climatological conditions. The other processes are: nivation deflation and the slowly creeping of the weathering mantle above the upper tree-limit and a little erosion of the streams, which are dissecting the rocky substrata only at the bars and the steps.

The described area, situated within the Tatra National Park, represents unique values of nature including geomorphology.

SUGGESTED READINGS

Armand, D., 1961: *Znaczenije gielomorfologii dla sielskogo chazjajstwa. Prikladnaja gieomorfologia*, Waprosy gieografii, 52.

Baszenina N., Gellert J., Joly F., Klimaszewski M., Scholz E., 1968: *Project of the unified key to the detailed geomorphological map of the world*, Folia Geographica, Ser. geogr.-phys., II Krakow.

Baszenina, N. i inni, 1971: *Legend to the international geomorphological map of Europa 1 :2500000*, Inst. Geogr. Cz. Ak. Sc., Brno.

Bulletin IGU, 1970, Vol. XXI, No. 2.

Dokumente und Informationen zur Schweizerischen Orts-Regional-und Landesplanung, Disp. nr 42, 1976.

Klimaszewski, M. (ed.) 1963 : *Problems of geomorphological mapping*, Geogr. Studies, 46, Warszawa.

Klimaszewski, M.; 1963: *Landform list end signs used in the detailed geomorphological map*, Problems of Geomorph. Mapping. Geogr. Studies, 46, Warszawa.

Klimaszewski, M., 1968: *Problems of the detailed geomorphological map*. Folia Geographica, ser, geogrphys., II, Krakow.

Klimek K., Kotarba A., Starkel B., Starkel, 1969, *Analiza i ocena srodowiska geograficznego powiatu ropczyckiego*, Dokum. Geogr. Inst. Geogr. PAN, 2/3.

Report of the Working Team for the Geomorphological Map of Europe on 1 : 500000, 1968, Inst. Geogr. Cz. Ak. Sc.

Scholz E., 1973, *Geomorphologische Karten und Legenden ausgewählter Massotabsgruppen*, Studia Geographica, 32, Brno.

St. Onge D., 1968, *Geomorphic maps, Encyclopedia of Geomorphology*, N. York.

Tricart 1965, *Principles et methodes de la geomorphologie*, Paris.

Ziemicki S., 1968, *Melioracje przeciwerozyjne*.

Carcenac C., Coincon R., Taillefer F., 1969, *Carte geomorphologique du Mas d'Azil SE. Rev. Geogr, d. Pyrenees et d. SO*, 40.

Castiglioni G., 1969: *L'attivita della commissione per la ricerca e la cartografia geomorphologica dell U.C.J.*

Literature Concerning the Geomorphological Mapping

Publications on the morphological mapping up to 1967 are listed in M. Klimaszewski, *Problems of the detailed geomorphological map*, Folia Geographica, II, 1968.

Publications on the geomorphological mapping up to 1971 are listed in: E. Scholz, *Geomorphologische Karten und Legenden. . . .* .Studia Geographica, 32, 1973.

Publications since 1971

Demek J., edit., 1971: *Manual of detailed geomorphological mapping*, Czechoslovak Akad, Sc., Brno.

Dorsser H. van, Salomé A., 1973, *Different methods of detailed geomorphological mapping (with coloured example)*, KNAG Geografisch Tijdschrift, 7.

Dorsser H. van, Saiomé A., 1974, *Two methods of detailed geomorphological mapping*, KNAG Geografisch Tijdschrift, 8.

Gellert J.; 1972, *Projekte und Probleme der internationalen geomorphologischen Forschung und Kartierung*, Peterm, Geogr. Mitt. 116.

Gobel P.; Leser H., Stáblein G., 1973, *Geomorphologische Kartierung. Richtlinien zur Herstellung Geomorphologischer Karten 1 : 25000.* Geom. Karte der Bundesrepublik Deutschland, Marburg.

Gohl D. 1972, *Strukturen und Skulpturen der Lanschaft. Die Methodik der Darstellung am Beispiel einer Karte von Deutschland*, Inaugural-Dissert., Bonn.

Hagedorn J., *Uber die Konzeption neuer geomorphologischer Karten kleinen Masstabs.* Wissensch. Redaktion IV.

Markovic M., 1973: *General approach to a geomorphological map with a model proposed for the geomorphological map of Yugoslavia*, Annales Geolog. de la Peninsule Balcanique, 38, Beograd.

Panizza M., 1972: *Schema di legenda per carte geomorphologische di dettaglio*, Boll. Soc. Geol. It. 91.

Pellegrini G., Savro W., Tessari F., 1972, *La recenti riunioni della commissione per la ricerca e la cartografia geomorfologica dell U. G. J.*, Boll. Soc. Geogr. Italiana.

Scholz E., 1973; *Geomorphologische Karten und Legenden ausgewahlter Massstabsgruppen*, Studia Geograph. 32, Brno.

St. Onge D., 1974: *The geomorphology of the Swan Hills area*, Alberta, Geol. Surv. of Canada, 74—25.

Tricart J. i in., 1972: *Cartografie Geomorphologique Traveaux de la RCP*, 77, Mam. et Docum., Ns 12, Paris.

Verstappen H., Zwidam R. van, 1968: *ITC System of geomorphological survey*, ITC Textbook of Photo-Interpretation, 7, Delft.

Verstappen H. 1970: *Introduction to the ITC system of geomorphological survey*, KNAG Geografisch Tijdschrift, IV.

THIRTEEN

Jaromir Demek

GEOMORPHOLOGICAL MAPPING—PROGRESS AND PROBLEMS

INTRODUCTION

IN THE last three decades geomorphological mapping achieved an important position in geomorphological research. Geomorphological mapping is nowadays the main research method of geomorphology in many countries. The geomorphological map is no more only a supplement of the text but an independent and indispensable result of geomorphological research. It is thus of great significance for the knowledge of relief of our planet on different levels since it reflects in a telling way the spatial inter-relations and associations of various relief elements. The compilation of a geomorphological map requires more detailed investigations of relief than the text, a system approach to the study of relief and the selection of information obtained from the viewpoint of the map purpose.

The heterogeneity of the Earth's relief, the differences in scales and the various demands on geomorphological maps caused a considerable variety of their contents and legends in various countries and even in the individual scientific centres of the different countries. Recently, after 1970, a new development of geomorphological mapping took place in connection with some changes in geomorphology as a science. These changes can be briefly characterized as follows:

(a) Better understanding of relief as the result of contradictory interactions of endogenic and exogenic factors.

owing to better rapprochement and interpenetration of geomorphology with geophysics and tectonics. New results of geophysical studies of the Earth's crust and the Earth's mantle, new geotectonic theories (such as for instance new global tectonics) and the development of methods of investigations of the Earth from the space resulted in the formation of a morphostructural tendency in geomorphology understanding better the relationship of relief to endogenic processes, mainly to those taking place in the upper mantle, and to fault tectonics.

(b) Better understanding of relief as one of the basic parts of the landscape sphere of our planet in which life and activities of human society take place; relief affects substantially the activities of man by its properties and the dynamics of various geomorphological processes; man affected on the contrary—accelerated or retarded—the course of most geomorphological processes influencing thus substantially the development of relief. These facts resulted in the origin of a new branch of geomorphology, the so-called ecological geomorphology.

(c) New methods of geomorphological research, mainly air and satellite photos, radar images and their processing by computers result in a considerable increase of the amount of information on relief and the possibility of fast cartographical elaboration. These facts support the development of constructive tendencies in geomorphology and the possibilities of geomorphological prognosis increasing simultaneously the practical significance of geomorphology mainly in the solution of one of the main problems of present-day mankind—the questions of complex care of environment.

New trends in geomorphology increased substantially the theoretical and practical significance of geomorphological maps. In this connection the interest in geomorphological mapping increased considerably manifesting itself in the higher number of publications dealing with geomorphological mapping and that of maps published in various countries (cf. D.V. Borisevich E. A. Lyubimceva 1971, L. R. Serebryany 1973). New handbooks devoted to methods of geomorphological mapping also appeared (A. I. Spiridonov 1970, 1975, I. P. Gerasimov 1970, J. F. Chemekove et al., 1972, N. V. Bashenina et al., 1975).

The IGU Commission on Geomorphological Survey and Mapping published in 1972 an international methodical handbook, the Manual of Detailed Geomorphological Mapping (J. Demek, ed. 1972) and in 1976 the Guide to Medium-Scale Geomorphological Mapping (J. Demek C. Embleton, eds. 1976).

In spite of this progress in geomorphological mapping there still exists a number of problems to be tackled. In the following text attention will be paid to some problems of geomorphological mapping on international scale as they follow from the activities of the IGU Commission on Geomorphological Survey and Mapping.

PROBLEMS OF THE CONTENT OF GEOMORPHOLOGICAL MAPS

The geomorphological map is a system of information representing a model of geomorphological geosystems existing really in relief.

Owing to the increasing amount of information on relief their recording on a single map becomes still more difficult. Therefore, even with geomorphological maps complexes develop (such as morphostructural, chronological, dynamic maps, etc.) similarly as in geology where lithological, stratigraphical, quaternary, tectonic, etc. maps exist. But I am nevertheless of the opinion that the conception of a general geomorphological map should be aimed at representing in a suitable selection of information the appearance, genesis and age of the relief. The general geomorphological map with this content is then a basis for partial, special and applied geomorphological maps.

On general geomorphological maps three categories of objects are usually represented in dependence on scale, such as:

(a) Parts of relief forms, i.e., elementary surfaces (sometimes called genetically homogeneous surfaces) and edges separating them.

(b) Relief forms, i.e., three-dimensional objects of various dimensions and complexity.

(c) Relief form associations, i.e., form complexes forming natural geomorphological complexes (relief types, morphostructures etc.).

Geomorphologists usually represent on detailed geomorphological maps (1 : 10000—1 : 100000) parts of relief forms, and on geomorphological maps on scales below 1 : 1 mill, relief form associations. Most problems occur with medium-scale maps on scales between 1 : 100000 and 1 : 1 mill. Some authors are of the opinion that relief forms and their parts can be represented even on medium-scale maps inclusive of maps on 1 : 1 mill. For instance in the USSR, the type legend for geomorphological maps on different scales compiled by G. S. Ganeshin-V. V. Soloviev-J. F. Chemekov has been accepted for geomorphological maps on 1 : 1 mill, prepared within the frame of the State Geological Map of the USSR on 1 : 1 mill. The base of the legend is the mapping of genetically homogeneous surfaces separated by edges (J. F. Chemekov et al., 1972, p. 245). On the Geomorphological map of the GDR on 1 : 750000 involved in the new National Atlas of the GDR relief forms will be represented.

But the IGU Commission on Geomorphological Survey and Mapping accepted after a longer discussion the recommendation that maps on scales from 1 : 100000 to 1 : 500000 be compiled as maps of relief forms and their parts and maps on scales between 1 : 500000 and 1 : 1 mill. as maps of relief form associations on the basis of morphostructures.

Maps of relief forms are indisputably more suitable mainly in practical application than those of form associations. But, on the contrary, they exhibit a number of disadvantages mainly if scales smaller than 1 : 500000 are concerned, such as:

(a) Maps of relief forms on scales smaller than 1 : 500000 are not lucid enough and become a mosaic of minute difficulty discernible areas; they represent insufficiently the appearance, mainly the plasticity of relief.

(b) In the effort for generalization joining of various forms and, accordingly, formation of form associations take place.

(c) Maps of relief forms express only difficulty neotectonic relief deformations mainly of megaanticline and megasyncline types.

The application of one legend for geomorphological maps on various scales is certainly attractive but the legends proposed so far do not meet all requirements. In the case of maps on scales smaller than 1 : 500000 maps of form associations are to be preferred.

PROBLEMS OF REPRESENTING GEOMORPHOLOGICAL OBJECTS

Geomorphological maps are a type of thematic maps and, therefore, governed by laws of thematic mapping. To represent geomorphological objects on geomorphological maps background colouring, conventional symbols and indices (characters, ciphers) are used. In background the fundamental tone and the tints are distinguished.

In thematic mapping the rule is valid that most efficient representation means are used for most important information involved in the map. In the case of general geomorphological maps most geomorphologists consider relief genesis most important. This is why colour is mostly used on general geomorphological maps to represent the genesis of geomorphological objects. But geomorphological maps are multi-layer maps and the colour tints must therefore express even another information of the content. Mostly two combinations occur, such as:

(a) Representation of the genesis of geomorphological objects by fundamental tone and their age by a tint of the fundamental tone (e.g. M. Klimaszewski 1968).

(b) Representation of the genesis of geomorphological objects by fundamental tone and of their inclination by a tint of the fundamental tone (e.g. J. Demek 1969). A.I. Spiridonov (1975, p. 13) mentions that the angle of slope can be estimated on the basis of contours which are an integral part of the topographical base of the geomorphological map. Our experience has shown that the contours are not sufficient to represent the plasticity of the relief and that even in the thematic content of geomorphological maps the slopes are to be distinguished according to their angle the inclinations having to be represented by tints of fundamental colours according to the principle the steeper the slope the darker the tint. The contours together with spot elevations allow to establish the absolute height position of the forms. But the general geomorphological maps must involve necessarily a classification of slope angles perceptible from the map at first sight. The IGU legend uses therefore the following classification of slopes: 0—2°, 2—5°, 5—15°, 15—35°, 35—55° and over 55°. This classification is not only of theoretical but even practical significance (cf. J. Demek, ed. 1972).

Another problem is the use of colours for genetic groups of geomorphological objects. For instance, in geomorphological mapping the forms are divided according to their genesis into glacial, fluvioglacial, cryogenic, marine, lacustrine, anthropogenic, etc. forms. Within these groups erosion-denudational and depositional forms are usually distinguished. The problem is if one fundamental colour is to be used for each genetic form group or if two colours are to be used separately for erosional and separately for depositional relief forms.

The IGU Commission on Geomorphological Survey and Mapping recommended to distinguish by fundamental colours erosion denudational and depositional forms for the following reasons:

(a) Erosion-denudational forms and depositional forms express two different tendencies in relief development in dependence on tectonic movements and climatic changes and their distinction reflects simultaneously the trend of geomorphological evolution

(b) The distinction increases the legibility of the map and allows to establish easier the chronology of the area represented

(c) The distinction improves the plasticity of the map from cartographical point of view.

A.I. Spiridonov (1975, p. 14) objects that this mode of division used in the IGU unified key to the detailed geomorphological map of the world (N.V. Beshenina-J.F. Gellert—F. Joly-M. Klimaszewski-E. Scholz 1968) results in the division in substance of some relief forms and elements and in the joining of forms of various genesis. He quotes as an example a talus and an accumulation river terrace which are represented by the same green colour and an accumulation and erosion river terraces represented by various colours (green and brown). In spite of this fact being a certain disadvantage the positive qualities mentioned in the previous paragraph prevail over the disadvantages.

The fact should be taken into consideration that in the construction of the legends to geomorphological maps the use of a high number of colour tints should be limited mainly for two reasons:

(a) better legibility of the map

(b) lower costs of map print which are rather high mainly in the case of systematic mapping affecting thus unfavourably the possibilities of the publication of maps.

In the legend mentioned 9 fundamental colours and 45 tints have been used. A cartographical analysis has shown that, for instance, in the case of small yellow areas the distinction is rather difficult.

The representation of bedrock and superficial deposits by colours on geomorphological maps used for instance usually on French geomorphological maps is unsuitable. French geomorphological school achieved great success in geomorphological mapping not only of France (cf. e.g. A. Journaux 1969, 1971, 1973, etc.) but even of various other regions of our planet. But certain specific relief properties of France and North Africa affected unsuitably the legend of French geomorphological maps. Lithology should not be neglected in any case on geomorphological maps because mainly erosion-denudational forms depend to various extent on the rocks of the bedrock (their composition, deposition, jointing, etc.). Even forms occur linked immediately to some rock types (e.g. karst forms). The IGU legend expresses therefore—mainly in the case of structural forms—the properties of the rocks of the bedrock and superficial deposits by special symbols.

Experience has shown that the representation of relief age by fundamental colour is less suitable. The establishment of the age of relief is one of most difficult problems in geomorphology. A rather wide space of time of the age of various forms mainly slopes (for instance Paleogene-Holocene) is to be quoted rather often. This fact makes the representation of the age by colour difficult. In addition, a small change of the age of forms established by further research debases the whole map. The IGU Commission on Geomorphological Survey and Mapping recommended therefore to express the age of the forms by indices (characters and ciphers).

THE NECESSITY OF STANDARDIZATION OF GEOMORPHOLOGICAL MAPS

In the present-day stage of development of geomorphological mapping on world-wide scale the necessity of standardization geomorphological maps on various scales comes still more to the fore. The geomorphological maps are nowadays acknowledged components of complexes of thematic maps for science and practice. They are still more applied in the solution of problems of national economy—in mineral prospecting, planning, protection against catastrophic phenomena (floods, landslides, etc.). But a further propagation of the practical application of geomorphological maps is impeded due to the heterogeneity of the content and means of representation of geomorphological maps in different countries or even in the individual geomorphological centres.

Geomorphologists become still more aware of this fact and, therefore, as early as in 1965 at the meeting of the IGU Commission on Applied Geomorphology held in Czechoslovakia work was started aiming at the standardization of geomorphological maps on worldwide scale. In 1965, the already mentioned projects of the Unified key to the detailed geomorphological map of the world (N.V. Bashenina-J.F. Gellert-F. Joly-M. Klimaszewski-E. Scholz 1968) appeared. This legend became the basis of further activities of the Commission on Geomorphological Survey and Mapping established at the 21st International Geographical Congress in India, 1968. Practical experience has shown that the legend is suitable for the compilation of detailed geomorphological maps on 1 : 10000—1 : 100000. The legend is to be understood as an outline which can be completed according to need in keeping the basic parameters. It is to serve general geomorphological maps and cannot therefore fully meet the demands of morphostructural or special maps (e.g. structural-geological oil prospecting).

The legend exhibits some insufficiencies. It has been for instance criticized by N.V. Bashenina (1975, p. 68) for not paying attention enough to forms of tropical clima-morphological regions, limiting of the representation of free and bound solifluction only to plains, insufficient expressing of slope processes. Critical comments were also submitted by A.I. Spiridonov (1975a, pp. 13—14 ; 1975b, pp. 24—26).

In spite of these imperfections the legend is a significant step towards standardization of geomorphological maps and was the basis for the publication of methodical international handbooks of detailed geomorphological mapping in English (1972), Russian (1976) and German (1976).

At the present time, the IGU Commission on Geomorphological Survey and Mapping is concerned with the standardization of medium-scale geomorphological maps. To the 23rd International Geographical Congress in Moscow the methodical handbook Guide to Medium-Scale Geomorphological Mapping will appear (J. Demek-C. Embleton eds. 1976) involving all international experience with medium-scale geomorphological maps. The Guide also comprises sample legends to general medium-scale geomorphological maps and examples of maps. With medium-scale geomorphological maps the question of standardization is more complicated than with detailed geomorphological maps. This is why the recommendation of the IGU Commission on Geomorphological Survey and Mapping is to be understood as an instruction for the compilation of regional legends without which we cannot obviously do so far.

INTERNATIONAL SMALL-SCALE GEOMORPHOLOGICAL MAPS

The progress in geomorphological mapping in recent years has created conditions for another new stage of development of small-scale geomorphological mapping. Most small-scale geomorphological maps (smaller than 1 : 1 mill.) have been compiled so far within national frames. The IGU Commission on National Atlases played an important part in their unification. But in spite of this fact, the analysis of small-scale geomorphological maps exhibits considerable differences in the content and means of representation of these maps. The differences make the publication of international small-scale geomorphological maps of groups of countries or continents considerably more difficult.

The compilation of international geomorphological maps is of considerable scientific and practical significance mainly for the following reasons:

(a) The compilation of small-scale geomorphological maps of larger territories requires the formation of scientific concept of relief development on a level satisfying the international scientific public; thus the compilation of the map results in a rapprochement of different opinions on international level, and a further evolution of the theory of geomorphology.

(b) The compilation of an international geomorphological map requires a uniform approach to the representation of the content which leads to the development of the theory of thematic mapping.

(c) The compilation of small-scale maps of larger territories allows to get acquainted with phenomena and laws unknown so far without a map.

(d) The compilation of international geomorphological maps requires a closer cooperation of scientists of various countries and often results in vivid discussions allowing to define the programme of work with more precision and leading to a more intensive international cooperation and mutual understanding

Within the frame of broad international cooperation the International Geomorphological Map of Europe on 1 : 2.5 mill, is being compiled. The idea of the compilation of this map originated in 1965 at the meeting of the IGU Commission on Applied Geomorphology held in Czechoslovakia. The original scale of 1 : 500000 has been found unreal. Owing to this fact the scale of 1 : 2.5 mill. was adopted after the establishment of the IGU Commission on Geomorphological Survey and Mapping in 1968. This scale is suitable for the following reasons:

(a) There exists a new topographical world map 1 : 2.5 mill.

compiled by the cartographical services of socialist countries.

(b) The map scale is detailed enough for a map of a continent.

(c) The map of Europe on this scale can appear not only in sheets but also as a wall map.

The original topographical base was elaborated by Kartografic Praha in a rectangular grid for the purpose of the compilation of the map of Europe.

After longer discussions a legend was elaborated for this map, whose 5th version is binding for the map of Europe mentioned.

At the present time, authors' originals of a considerable part of Western and Central Europe, the Ural and the northern section of the European part of the USSR are available. Unfortunately, difficulties appeared mainly in Western Europe as to contacting the various morphostructures and forms (e.g., moraines) at the frontiers of individual countries. But simultaneously, the cooperation among European geomorphologists improved and their willingness to a rapprochement of their different viewpoints increased.

The print of the sample sheet was delayed owing to the tardy delivery of the topographical base. But now, the sample sheet of a complicated territory of Central Europe is in the press and will be submitted at the 23rd International Geographical Congress in the USSR in this year.

In the course of the preparation of the International Geomorphological Map of Europe many difficulties will arise because it is necessary to compile geomorphological maps of geomorphologically relatively less investigated territories of Europe and North Africa where geomorphological maps have been compiled so far only in small measure. On the other side, the work on the map resulted in an unprecedented activity of European geomorphologists and valuable experience was gained as to international coordination of work.

A significant contribution to the realization of this work has been UNESCO's permanent support. Dr. K.O. Lange, UNESCO's representative, takes active part in the meetings of the IGU Commission on Geomorphological Survey and Mapping.

In the next period, the IGU Commission on Geomorphological Survey and Mapping intends to elaborate a regional geomorphology of Europe on the basis of experience with the compilation of the International Geomorphological Map of Europe.

CONCLUSIONS

The unprecedented development of geomorphological mapping after World War II means a new quantitatively and qualitatively superior stage in the development of geomorphology as a science. In the last three decades geomorphologists gained considerable experience and passed successively over to the standardization of geomorphological maps on various scales within international frame. In the work on the International Geomorphological Map of Europe a progress has been achieved. It is the first attempt at the unification of small-scale geomorphological maps on continent scale. The compilation of the map results even in a unification of opinions concerning the origin and development of relief in Europe.

But this positive development is not without problems. Several questions were discussed in this paper. But it should be realised that no further progress in geomorphological mapping and, accordingly, in geomorphology as a science is possible at the present time without the standardization of the content and modes of representation of geomorphological maps on various scales on international level.

SUGGESTED READINGS

Bashenina, V.V., 1974: O nekotorych shkolakh geomorfologicheskogo kartografirovanya za rubezhom. Vestnik Moskovskogo universiteta, *serya V geografya*, 28, 5, 28-34, Moskva.

Bashenina, N.V., 1975: O. mezhdunarodnoi legende dlia kart krupnykh masshtabov, Bashenina, N.V., ed., *Geomorfologicheskoie kartografirovanye v. s'emochnykh masshtabakh*. Izdatelstvo moskovskogo universiteta, 18-68, Moskva.

Bashenina, N.V., (editor), 1975: *Geomorfologicheskoie kartografirovanye v s'emocenykh masshtabakh.* Izdatelstvo moskovskogo universiteta, pp. 264, Moskva.

Bashenina, N.V., Gellert, J.F., Joly, F., Klimaszewski, M., Scholz, E., 1968: Project of the unified key to the detailed geomorphological map of the world. *Folia Geographica* II, serie geographica-physica, Polska Akademia Nauk, Krakow.

Borisevich, D.V., Ljubimceva, E.A., 1971: Geomorfologicheskoie kartirovanye. Itogi nauki, *Geomorfologya*, 2, 1-149, Moskva.

Chemekov, J.F. et al., 1972: Metodicheskote rukovodstvo po geomorfologicheskim issledovanyam. *Nedra*, pp 384 Leningrad.

Demek, J., 1967: Generalization of geomorphological maps. Progress made in geomorphological mapping. Geograficky ustav CSAV v Brne. *Zpravy o vedeckych vyzkumech*, 9, 36-72, Brno.

Demek, J., 1969: Podrobna obecna geomorfologicka mapa 1:25,000 (list Dolni Kounice). Detailed general geomorphological map 1 : 25000 (Dolni Kounice Sheet). *Studia Geographica* 1, 139-148. Ceskoslovenska akademie ved—Geograficky ustav Brno.

Demek, J., 1970: Proposal of the International Geomorphological Map of Europe at scale 1 : 2500000. 21st International Geographical Congress India 1968, Papers 1, 28-29. National Committee for Geography, Calcutta.

Demek, J., 1972: International geomorphological map of Europe on 1 : 2500000. *Sbornik Ceskoslovenske spolecnosti zemepisne* 77, 2, 21-127, Praha.

Demek, J., 1975: Mezhdunarodnye tematicheskye karty. International thematic maps. In: A.M. Berliant—A.V. Vostokova, ed., *Puti razvitya kartografii Izdatelstvo moskovskogo universiteta*, pp. 210-216, Moskva.

Demek, J. (editor), 1972: *Manual of detailed geomorphological mapping*, pp. 344, Academia, Praha.

Demek, J., ed., 1974: Problems of medium-scale geomorphological mapping. *Studia Geographica*, *41*, 1-206. Ceskoslovenska akademia ved—Geograficky ustav Brno.

Demek, J., Embleton, C., 1976: Guide to Medium-scale Geomorphological Mapping. *Academia*, Praha.

Ganeshin, G.S., ed., 1975: Problemy geomorfologicheskogo kartirovanya. Pezisy dokladov Vsesoyuznogo soveshchanya po raznomasshtabnomu geomorfologicheskomu kartirovanyu 12 maja 1975 g. VSEGEI—Institut geografii AN SSSR, pp. 232, Leningrad.

Gellert, J.F., 1968: Das System der komplex geomorphologischen Karten. *Petermanns Geographische Mitteilungen 112/3*, 185—190, Gotha.

Gellert, J.F., 1969: Die Konzeption und unifizierte internationale Legende fur geomorphologische Ubersichtskarten europalscher Lander. *Geographische Berichte*, *14*, 2, 132-35, Berlin.

Gellert, J.F., 1970: The System of Complex Geomorphological Maps. 21st International Geographical Congress India 1968, Papers 1, 63-65. *National Committee for Geography*, Calcutta.

Gellert, J.F., 1971: Internationale Konzeption und unifizierte Legenden fur geomorphologische Karten verschiedener Masstabe Wissenschaftliche Zeitschrift der Padagogischen Hochschule Potsdam, pp. 423-429. Potsdam.

Gellert, J.E., 1972: Projekte und Probleme der internationalen geomorphologischen Forschung und Kartierung. *Petermanns Geographische Mitteilungen. 116*, 1, 75-79, Gotha.

Gellert, J.F., 1974: Probleme einer Anleitung zur mittelmasstabigen geomorphologischen Kartierung. *Petermanns Geographische Mitteilungen 118*, 2, 153-156, Gotha.

Gerasimov, I.P., editor, 1970: Primenenie geomorfologicheskikh metodov v strukturno geologicheskikh issledovanyakh Nedra, pp. 296 Moskva.

Gllewska, S. 1967: Different methods of showing the relief on the detailed geomorphological maps. *Zeitschrift fur Geomorphologie* NE, 11, 4, 481-490, Berlin W.

Journaux, A., 1973: Cartes des formations superficielles et cartes geomorphologiques de Basse-Narmandie au 1:50,000e (Feuille de Bayeux-Courseulles). *C.N.R.S. Centre de Geomorphologique Bulletin*, 17, Caen.

Journaux, A., 1971: Carte de formations superficielles et cartes geomorphologiques de Basse-Normandie au 1:5`000e (Feuille de Caen). *C.N.R.S. Centre de Geomorphologique Bulletin*, 11, Caen.

Journaux, A., 1969: Presentation des cartes des formations superficielles et des cartes geomorphologiques de Basse-Normandie au 1:50000e (Feuille de Medizon). *C.N.R.S. Centre de Geomorphologique Bulletin* 4, Caen.

Kugler, H., 1965: Aufgabe, Grundsatze und methodische Wege fur grossmassstabiges geomorphologisches Kartieren. *Petermanns Geographische Mitteilungen*, 109, 241-257, Gotha.

Kugler, 1968: Einheitliche Gestaltungsprinzipien und Generalisierungswege bei der Schaffung geomorphologischer Karten verschiedener Massstabe. In: H. Barthel, ed., *Landforschung Erganzungsheht*, 271, 259-279. Gotha-Leipzig.

Leser, H., 1972: Inhalt und Form ais Problem gross- und kleinmasstabiger geomorphologischer Karten. *Kartographische Nachrichten*, 22, 4, 156-165.

Serebriannyi, L.R., 1973: Geomorfologicheskoie kartovedenye *SSSR a chastel sveta*. Nauka, pp. 248, Moskva.

Scholz, E. 1973: Geomorphologische Karten und Legenden ausgewahlter Massstabe. *Studia Geographica*, 32, 1-120, Ceskcslovenska akademie ved – Geograficky ustav Brno.

Spiridonov, A.I., 1970: Osnovy obshchei metcdiki polevykh geomorfologicheskikh issledovanly i geomorfologicheskogo kartografirovanya. *Vyshshaya shkola*, pp. 456, Moskva.

Spiridonov, A.I., 1975a: O principakh postroyenya krupnomasshtabnykh geomorfologicheskikh kart. In: G.S. Ganeshin, ed. *Problemy geomorfologicheskego kartirovanya*. pp. 13-14, Leningrad.

Spiridonov, A.I. 1975b: *Geomorfologicheskye kartigrafovanya*. Nedra, pp. 183, Moskva.

Verstappen, H. Th., 1970. Introduction to the ITC system of geomorphological survey. *K.N.A.G. Geografisch Tijdscrift*, 4, 1, pp. 85-91, Amsterdam.

FOURTEEN

Johannes F. Gellert

PRINCIPLES OF GEOMORPHOLOGICAL REGIONALIZATION AND ENCODING OF GEOMORPHOLOGICAL CHARACTERISTICS FOR COMPUTER PROCESSING OF DATA

GEOMORPHOLOGICAL regionalization involves the delimitation, characterization and cartographical representation of regional units composed of similar structure of surface forms and forming regional units of various rank and size orders. It rests on geomorphological analysis of relief forms and relief forming processes and on the cartographical representation of the results of such analyses on detailed or even medium scale geomorphological map. It is based on the classification of relief forms in morphogenetic form groups as tectonic surface forms, fluvial, glacial, aeolian surface forms a.s.o (cf. Manual 1972, Gellert 1972, 1980) and on investigations of the regional distribution of these basic geomorphological units and their relations to higher-order spatial units. These geomorphological regionalization promotes the regionalization of other landscape elements as soil, watering, weathering and vegetation including summary nature resources and nature potentials in frame of a complex geoecological survey. It includes many hydrological and engineer-geological facts. Geomorphological regionalization in these manner is relevant in the fields of planning and economic development.

The basic unit in geomorphological regionalization is the morphotop or morphofacies (in PECSI and SOMOGY's sense 1968). These are geomorphological regional units characterized by :

—uniform shape (shape: morphography, morphometry)

—homogeneous geological especially lithological structure
—uniform origin and development (morphogenesis, morphochronology)
—uniform present-day processes (morphodynamics).

Their area may range from several square decametres (10 m^2) to any hectares (100 m^2). In exceptional cases of geomorphological uniformity, the units may obtain the extent and order of a geomorphological type group. These basic units of geomorphological regionalization comprise individuals forms or complexes of forms—concave, convex or flat—consisting of undecomposable form elements. Their representation to scale exceeds the possibility of the complex detailed geomorphological map 1 : 50000 and is accordingly the task of geomorphological plans on 1 : 10000 or detailed geomorphological maps on 1 : 25000.

The morphotops or morphofacies never occur as isolated forms but they also forms together joined regional units in the form of form complexes and form groups with similar but heterogeneous geomorphological marks of orographical, morphographical, morphometric, lithologico-sedimentological, morphogenetic-morphochronological and morphostructural kind. In sense of a sequence of topologic, chorologic, regionic and geospheric (global) dimension (in sense of the GDR geographers and other) the hierarchy of geomorphological regional units is built up as follows :

MT morpho-tops (morpho-facies)
MTG or MNCh groups of morphotops or morpho-nannoechores
MMiCh morpho-microchores
MMeCh morpho-mesochores
MMaCh morpho-macrochores

They are followed by units of higher order (morpho-regions) and of continental size order. In that sense the Central European Lowland (German-Polish-Lowland) is a microregion and Europe or southern Africa a macroregion, for example.

In comparison with morphotops the higher-order geomorphological regional units mentioned are—with exception of their area size and their rank and size orders (dimensions)—independent geomorphological units governed by laws and characterized by the various series but heterogeneous geomorphological marks. Whereas the morphotop groups consist of a mosaic of several morphotops the morphochores are noted for marks of various character but the same level, such as

—type relationship of geomorphological units situated in the system by one degree lower and forming a morphotop group
—general features of one or several geomorphological elements specific for all regional units
—morphogenetic appurtenance of several geomorphological regional units of lower order
—interlocking of various geomorphological regional units of lower order
—common position of various geomorphological regional units of lower order

The delimitation of these units is carried out on detailed geomorphological maps of about 1 : 50000 and medium scale geomorphological maps at about 1 : 200000 a.s.o. (Guide 1978). The areas of morphotop groups should be marked off on transparent overlay to the according geomorphological map by single, coloured, thin or interrupts lines, and for the delimitation of areas of higher unit order, coloured or interrupted lines should be used. In cases where the borders of units merge, the borders of the lower-order units are omitted in favour of the borders of higher-order units.

The individual regional units are, if possible, marked by local names indicating the geomorphological character, as well as by numbering according to a decimal system (1.1.1—see later), the last but one number signifying the morphotop group but the last number should be reserved for special subdivision into morphotops an geomorphological basic maps at 1 : 25000 and smaller scales. Every geomorphological regional unit obtains its numbering proper; the morphotops occurring in close proximity can be marked by the same number within the group of morphotops as area types in the same situation. So that the mesachores stand out of the maps, their indices should be shown in bold type. The explanations of the values of places in numbers should be given on the maps of geomorphological regionalisation as transparent overlay to the adequate geomorphological maps or special maps in the same cartographic basis and in text explanations to complex geomorphological maps.

Such a system of numbering of the geomorphological regional units, starting from the smallest units, the morphotops and groups of morphotops to greater of higher order, enables recording of geomorphological regional units of anything geomorphological map in a catalogue and within its scope the explanation of the delimited geomorphological regional units by the statement of their special marks and data which are of importance for geomorphological science and the application of its knowledge in other disciplines and, above all, in technico-economico practice. The data can be rapidly utilized, if necessary, in such practical applications as planning and economic development f.i. in optimal localization of agriculture and forestry, of constructure of industries and settlements and of routes on basis of geomorphological facts.

The hierarchical numbering system also facilitates matrix analysis and mechanical processes of the data. The matrices for mechanical data processing of geomorphological marks and indices should comprise—according to the respective indices for geographical position of the respective geomorphological regional unit (f.i. morphotop group) in the Earth grid and the sheet section of the respective geomorphological map—the following geomorphological marks:

—index of the geomorphological regional unit (morphotop and morphotop group and unit of higher order)
—area of the respective geomorphological regional unit in km^2
—extreme difference in altitude in m
—prevailing relief amplitude in m in periphery of 1 km
—density of breaks of slope (valleys, rills, steps) in length of breaks of slope in km/km^2

—prevailing slope class :

0	—	0°30′	plain
0°30′	—	2°	slightly sloping
2°	—	5°	gently inclined
5°	—	15°	strongly inclined
15°	—	35°	steep
35°	—	55°	very steep
more than 55°			precipitous

—prevailing substratum (solid rocks as magmatites, sedimentites and regionally metamorphozed rocks—loose rocks of various kinds and origins inclusive anthropogenic substratum of civil engineering kind)
—prevailing genetic form group characterized by the form creating processes (tectonic, volcanic, denudative, fluvial a.s.o.)
—prevailing thickness of weathering and debris covers over solid or loose rocks in x, xm
—lithological kind of weathered and debris cover
—humidity conditions of the debris cover controlling the denudation-erosion processes
—characteristic of the prevailing recently form controlling processes
—special potential form creating processes (rockfall, landslide, mud stream, flood and other)
—destruction of soil (soil erosion)

According to the number of columns available and the need, the number of geomorphological marks and characteristic data involved in the matrix can be increased, it may be necessary or useful in certain circumstances to add data from engineering-geological, pedological or hydrogeological and hydrological sources.

Marks and characteristic data of superior geomorphological regional units—morphochores a.s.o.—will follow from the complex registrations of geomorphological marks and characteristic data of the morphotop groups involved in the matrix as a resultant of an inward or mechanical integration. In regions mapped geomorphologically on a smaller scale, perhaps on 1 : 100000 or less, it will be possible to eliminate the morphotops. The geomorphological regionalization will be based more reasonably on morpho-microchores requiring for their characterization another complex system of marks and characteristic data based on the complex of homogenities and heterogenities of the chores.

By computer processing of coded geomorphological data in the described manner great economies in time and labour are possible. The computer can print out directly, for example, maps showing the distribution of any selected element or elements, or it can quickly assemble quantitative data as necessary. Specific geomorphological data can rapidly be extracted from complex geomorphological detailed or medium-scale maps. The codification of the content of detailed or medium-scale geomorphological map compiled on the basis of original field research thus enables a reliable, complete and versatile evolution for many scientific and technico-economic purposes. It represents a rationalization in the application of the results of extensive basic research.

SUGGESTED READINGS

Gellert, J.F., 1972: Geomorphological regionalization, In: *Manual of detailed geomorphological mapping*, Prague, pp. 130-141.

Gellert, J.F., 1980: A uniform system of the surface forms morphogenesis and geomorphological classification and regionalization.

Guide to medium scale geomorphological mapping. Brno 1978.

Pecsi, M.; S. Somogy 1968: Subdivision and classification of the Physiographic landscapes and geomorphological regions of Hungary. *Research Problems in Hungarian applied geography*. 24 Budapest, p. 7.

Manual of detailed geomorphological mapping. Prague, 1972.

FIFTEEN

Johannes F. Gellert

A UNIFORM SYSTEM OF THE SURFACE FORMS MORPHOGENESIS AND GEOMORPHOLOGICAL CLASSIFICATION AND REGIONALIZATION

THE SUBMITTED "System of the Surface Forms Morphogenesis and Geomorphological Classification and Regionalization" (Fig 15.1) is based on the widely recognized knowledge that the formation of the Earth surface forms, the morphogenesis, occurs due to the dialectic alternate effects realized within the sphere of the Earth surface with the effects or overcoming of the gravitation, these effects occurring between endogenous and exogenous processes caused by the telluric processes of the development of the material in the Earth body as they are investigated by geophysics, geochemistry and geotectonics, and by the insolation of the planet Earth by the sun and the processes evoked in the atmosphere in this way. In this sense the exogenous processes, such as weathering, denudation (destruction) and accumulation as well as the destruction and accumulations forms created by them, are conditioned by atmospheric components, mainly temperature and precipitation and controlled by the global regional differentiation of weather and climate. The destruction and accumulation forms created by the differentiated exogenous processes determine the morphosculpture of the Earth regions.

In the endogenous sphere tectonic and magmatic processes are to be distinguished within scope of their mutual dialectic relations. They determine the morphostructure of various regions (I.P. Gerasimov 1947, I.P. Gerasimov-J.A. Meshcheryakov 1967). The magmatic processes share, in form of a recent and sub-recent neovolcanism with the development of

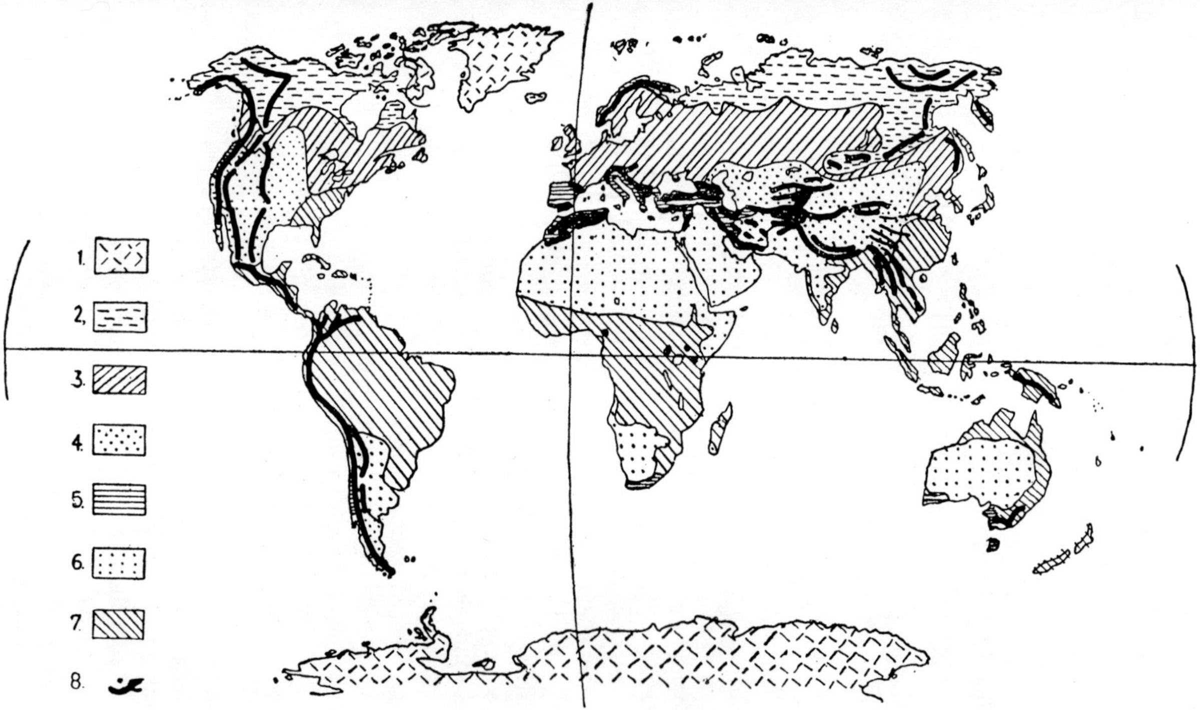

FIG. 15.1 : *Scheme map of the zonal spheres of the present morphological processes*

Legends—Zonal areas: 1. of inland-ice 2. of perennial frost and permafrost 3. of perennial precipitation with mostly moderate temperatures and/or not frost 4. of intra-continental dry regions and/or not frost and/or not snow cover 5. of summer warm, with winter rainfall and mostly periodic rivers 6. of subtropical and tropical dry areas 7. of subtropical and tropical warm moist regions with well or not well developed dry season, partly periodical and perennial rivers 8. of Extra-zonal areas of mountains and high lands with frost and snow.

volcanic accumulation and eruption forms, actively in the modelling of the Earth surface forms. With the texture and structure of the magmatites (plutonites and volcanites) and their resistance to exogenous processes the magmatic processes altogether (neovolcanism, paleovolcanism, plutonism) condition passively the formation of structural forms.

The tectonic processes divided according the geological age into neotectonics, involving neogene and quaternary, and paleotectonics allow the differentiation into epeirogenic, dictyogenic and tectogenic (marked till the present mostly but wrongly as orogenic) types of movement of the Earth crust to which correspond the structure type of the passive platforms (blocks, shields, stabile shelves), activized platforms (labile shelves) and the young tectogens (folded systems designated till present wrongly as orogenes) of the sphere of which the predominantly weak, moderate or strong young crust movements (neotectonics, morphotectonics) are characteristic. The structure of the Earth crust created within the scope of paleotectonics and neotectonics discernible in the deposition of the rocks (structure, tectonics) develop passively in the relief in the form of structural forms due to the lithologically conditioned different rock resistance to exogenous processes. They form together with the structural forms conditioned in the same way by the different resistance of the magmatic rocks to exogenous processes the passive morphostructure of the Earth relief.

Contrary to the structure tectonics passing passively into the relief can be put the morphotectonics and the recent tectonics within the scope of which young crust movements play an active part in the modelling of the Earth surface. They create mainly the relief energy and determine together with the volcanic forms the active morphostructure of the Earth relief. Here appears as necessary the excorporation of the morphotectonics creating actively surface forms since, on the one hand, the neotectonics involve also structure forming processes their course being of no direct significance in the formation of the present surface forms, and on the other hand, crust movements acting in the Pre-Neogene period and being thus not involved into the term neotectonics took part in the modelling of the present relief. For similar reasons the recent tectonics should be also excluded of the scope of the morphotectonics because the recent crust movements in form of earthquake or earth movements demonstrable but fine levelling are but of a subordinate, in substance modelling, significance within the scope of the origin and development of surface forms determining the relief at present.

The origin and formation of the surface forms (morphogenesis) occurring in the described conditions and dialectic alternate effects enable a multiple classification of the Earth surface forms. Each of those classifications is determined through the importance attributed to certain agents of the morphogenesis within the scope of their mutual dialectic relations. These various surface forms classifications do not eliminate mutually but they are in a close dialectic relation being superimposed in conformity with the lows what enables their combination in geomorphological representation.

According to form creating processes those form groups can be characterized first in which Earth surface forms are brought together into genetic groups of forms according to the morphogenetic points of view. On the basis of the recent morphological processes controlled by the present weather and climatic conditions and producing mostly but a modelling effect on the surface forms developed under different earlier weather and climatic conditions (paleoclimates) can be characterized all over the Earth the morphodynamic form associations in zonal and azonal spheres (see Table 15.1). Finally, a natural classification of the Earth surface according the macroform types on morphostructural basis can be observed within the scope of which a destruction relief on the basis of the above mentioned tectonic structural and neotectonic movement types, and an accumulation relief on the basis of genetic form groups and the morphostructural or morphosculptural character of its sphere are distinguished.

Outside the morphogenetic principle represented here there is a morphographic-morphometric classification of the relief and its surface forms according the relief classes as they are defined by the relative relief energy and the slope conditions and according to form classes, i.e., the size of forms superimposed or integrating in the relief and resulting from uniform morphological processes.

TABLE 15.1: **Morphodynamic form associations by the present morphological processes in zonal spheres**

---

zonal areas
- of inland-ice
  - (subglacial exaration, glacifluvial erosion)
- of perennial frost and permafrost
  - (frost weathering, cryoturbation, congelifluction, cryoplanation)
  - high polar frost talus zone
    - (extreme frost weathering, cryoturbation, free congelifluction, fluvial erosion, intensive valley forming and fluvial accumulation)
  - tundra zone
    - (bound congelifluction, periodic wash, forming of dells and valley floors)
  - permafrost zone
    - (thermocarstic processes, fluvial erosion. forming of peat bogs)
- of perennial precipitation with mostly moderate temperatures ± frost
  - (chemical weathering, subcutan creeping, moderate fluvial erosion)
  - zone with perennial or summer precipitation, perennial rivers, regular frost and snow cover and permafrost isle-likely
    - (frost weathering, periodic fluvial erosion, forming of pest bogs)
  - zone with perennial or summer precipitation, perennial rivers, regular frost and snow cover without permafrost
    - (frost weathering, perennial fluvial erosion)
  - zone with perennial or summer precipitation and perennial rivers
    - (perennial fluvial erosion)
- of intra-continental dry regions ± frost and ± snow cover
  - (mechanical weathering, eolic deflation and accumulation, episodical torrential fluvial erosion)
  - zone of intra-continental winter cold steppes regions with summer precipitation and periodical and perennial rivers and dust storms
    - (periodical and perennial fluvial erosion)
  - zone of intra-continental deserts and semideserts of middle latitudes with high summer temperatures and ± frost, periodical and episodical rivers and allochton streams
    - (periodical and episodical fluvial erosion)
- of summer warm, with winter rainfall, mostly periodic rivers
  - (chemical weathering, mainly periodical fluvial erosion, slopes wash)
- of subtropical-tropical dry areas
  - (mechanical weathering, eolic deflation and accumulation, pediment and glacis processes)

TABLE 15.1 *contd.*

zone of subtropical-tropical warm semiarid dry regions with small winter rainfall in few months
(periodical or episodical slope and sheet wash and fluvial erosion, crust forming)

zone of tropical heath full arid dry regions with great temperature amplitudes, fewer than 100 mm rainfall, mostly episodical rivers and any allochton rivers
(eolic corrasion, episodical slope and sheet wash and fluvial erosion)

zone of tropical-heath, semiarid dry regions with long dry season and summer rain season, periodical rivers and any allochton streams
(periodical slope and sheet wash, fluvial erosion and crust forming)

of subtropical-tropical warm moist regions ± well developed dry season, partly periodical and perennial rivers
(deep going chemical weathering ± crust forming)

zone of tropical heat, seasonal moist semihumid region with well developed dry season of 3-6 arid months with periodical and perennial rivers
(mechanical weathering, well developed slope and sheet wash, peneplain/etchplain—and inselberg-processes)

zone of tropical seasonal moist semihumid temperated high land-areas with rainfalls in 7-10 humid months and periodical-perennial rivers
(subcutan solifluction)

zone of subtropical heat moist semihumid regions with ± developed dry season and (trade-wind or monsoon) rainfall with p riodical rivers and dry perennial allochton rivers
(subcutan solifluction, suffosion, landslides, slope wash, fluvial erosion, tropical carst forming processes)

zone of tropical heat humid-semihumid regions with well developed dry season with ± mostly perennial rivers
(subcutan solifluction, suffosion, landslides, fluvial erosion, tropical carstic processes)

extrazonales areas of mountains and high lands with frost and snow
(frost weathering, cryoturbation, congelifluction)

of mountains with regular or permanent frost and firn and ice (glaciers)
(subglacial exaration)

of high lands with regular frost and ± snow and strong pushed drainage
(periodical fiuvial erosion and accumulation, eolic deflation and accumulation)

a zonal area of inundation of the great streams and lacustrin lowlands
(alluvial accumulation)

In the nature all that surface forms are incorporated in regional form complexes according a hierarchic system of geomorphological areas similar to the landscape areas. Analogue to that there are geomorphological areas in different dimensions (scales). According to the terms of the geographers (geoecolists) of the G.D.R. and others there are groups of morphotops, morphochores and morphoregions with subdegrees nanno-, micro-, meso-, macro- and megachores-regions respectively. That geomorphological regionalization of the Earth relief (Georelief) with its differentioned surface forms and morphogenesis in process and time (morphochronological) in texts, computer datas and maps is the basis and the aim of a complete and diverse scientific knowledge and reproduction (documentation) of the Earth surface forms by the geomorphology as a geographical or geological discipline in the frame of geosciences.

SUGGESTED READINGS

Bubnoff, S.V., 1938: Uber die Gerustbildung der Erdrinde (Dictyogenese. *Die Naturwissenschaften*, pp. 745-755.

Gellert, J.F., 1969: The system of the Morphogenesis and Morphogenetic Classification of the earth surface forms as the basis of the conception of geomorphological maps, *Studia Geographica Brno*, pp. 32-36.

Gellert, J.F., 1972: Geomorphological regionalization, *Manual of Detailed Geomorphological Mapping*. Prague, pp. 130-150.

Gellert, J.F., 1980: Geomorphologische Erdkarten und weitere Probleme einer globalen Geomorphologie, *Studia Geographica Brno*. (in print).

Gerasimov, I.P., 1947: Opyt geomorfologiceskoj interpretacii obsceji schemy geologiceskogo strojanija SSSR (Test of geomorphological interpretation of a general scheme of geological structure). *Problemy fiz-geogr.* 12, Moscow.

Gerasimov, I.P., J.A. Mescerjakov, 1967: *Rel'ef Zemli* (Relief of the Earth). Moscow.

SIXTEEN

Naresh Bedi

APPLIED GEOMORPHIC STUDIES ON LANDSAT IMAGERY

INTRODUCTION

AEROSPACE remote sensing techniques have opened new vistas for landscape analysis and planned development of regional economic resources (Verstappen, 1970; K.R. Rao, 1978; Bedi, 1979). These techniques have reversed the earlier system of construing regional picture on the basis of the extrapolation of local information. Now it is possible to study and analyse the large areas directly on aerospace imagery. The geomorphic analysis on them culminates into breakdown of the terrain or region into individual landforms and related processes which can be later evaluated for the economic potential. The classification of landscape with keeping in view their genetico-chronological aspects makes it possible to separate active to non-active forms (St. Onge in Fairbridge, 1967). When such information is put on the geomorphic maps they become extremely useful to the planners. Specific theme geomorphic maps can be easily prepared from space imagery like Landsat or aerial-photographs supplemented by nominal field work. The importance of such geomorphic studies is given in Table 16.1.

Suitability of Remote Sensing data for Applied Geomorphic Studies

In the present scenario of alround development, there is an urgent need for speedy acquisition of new information and

TABLE 16.1 : **Application of remote sensing-geomorphic mapping in planning and economic development**

1.	Land use	Territorial planning Regional area planning Conservation of natural and cultural landscape.	
2.	Agriculture and Forestry	Soil Conservation Soil erosion control Soil reclamation Potential utilisation Drainage and Irrigation Land Facet study	
		Communications	Roads Railways Canals Harbours-Site Alignments
3.	Civil Engineering	Resettlements Hydro-engineering	Dams and Reservoirs Regulations of rivers Canals Shore Protection Harbour construction
4.	Geology and Mineral Exploration		
5.	Hydrogeomorphology—ground water exploration.		

(Modified after Demeck 1972)

updating the existing one. This requirement can be very well fulfilled by the use of aerospace remote sensing data for geomorphic studies. The remote sensing imagery by its capability of providing synoptic view of a large area (32,900 sq. km. in one Landsat picture) helps in getting a regional picture at a glance. It also facilitates the direct comparison of landscape as well as geomorphic processes going on in otherwise widely separated areas. The possibility of getting temporal data over the same area periodically helps in monitoring the changes taking place in an area with time, e.g., coastal erosion, accelerated erosion in an area, fluvial changes, desertification, etc. In other words natural activities or processes which have direct effect over the human welfare can be monitored. The areas which are either inaccessible or have hampered mobility can be directly studied on the aerospace imagery in less time and cost.

The Landsat Multispectral Scanner (MSS) records data in four bands of electromagnetic spectrum (0.5 to 1.1 μm) which makes it a highly versatile form of data. It can be used to derive information about most of the landscape factors simultaneously. As the inter-relations of the various landscape ecological factors such as landforms, soil, hydrological conditions, land cover etc. are reflected in the terrain and in the imagery by a degree of correlation in the distribution patterns of these components (Verstappen, 1977). Thus geomorphological phenomena when studied in an environmental context yield information about other components of the landscape. For a quick appraisal of landscape elements in undeveloped areas and detection of hazards use of aerospace remote sensing data and techniques is highly desirable. Two examples of applied geomorphic studies carried out on Landsat imagery are given here.

EXAMPLE : 1

Hydrogeomorphic Study in Semi-arid area in Central part of Andhra Pradesh (India)

Area Background

The study area is bound by latitudes 15° 00′ 00″ and 16° 30′ 00″ N and longitudes 78° 00′ 00″ and 79° 30′ 00″ approximately. Most parts of the area need ground water development. Geological, geomorphological and climatic conditions are highly variable. About 90% of the area is occupied by hard rocks only. Most of the area experiences a semi-arid type of climate where rainfall ranges between 620 and 890 mm with an inconsistent distribution (Plate 1).

Approach

In semi-arid and arid areas ground water is perhaps the most important source of fresh water for most part of the year. As a function of the hydrological cycle most of the water originates at the surface through precipitation which subsequently percolates subsurface to form the ground water. During this transpiration also takes place, which in one way or the other, effects the subsurface permeation of water. Two groups of factors interact during this process and they are:

1. Hydrometeorological elements—mainly precipitation
2. Terrestrial elements—landform and landscape elements.

Since combinations of these elements vary from area to area therefore the generation, mobility, distribution and stability of ground water is a manifestation of such combinations.

Amongst terrestrial elements the geological base is an independent variable. The landforms and landcover are controlled by geological base and microclimatic conditions. Therefore, hydromorphological analysis involves delineation of landforms and their land cover associations. Later, they can be evaluated for their role in the amelioration of ground water.

Methodology

Following forms of Landsat data were used as data inputs and interpreted visually to decipher and demarcate the above mentioned terrestrial elements.

1. Landsat Scene (Path-Row No. 154—049) dt. 28th Feb. 1975—Black and white prints of bands 5 and 7 on scales 1 : 1 million and 1 : 250,000 (Plate 1)
2. False colour composite on scale 1 : 1 million.

The related ancilliary information was corroborated from the corresponding Survey of India topo sheets on 1 : 250,000 scale and Geological Survey of India geological map of Andhra Pradesh on 1 : 1 m scale. The map prepared was partially field checked and corrected.

Geological Set-up

The area is occupied by a variegated assemblage of rock types. Granite-gnesses (GR) with dykes (Archaean) occupy the north-western and western parts. There are several sets of lineaments both on micro and mega scales (Fig. 16.1). The

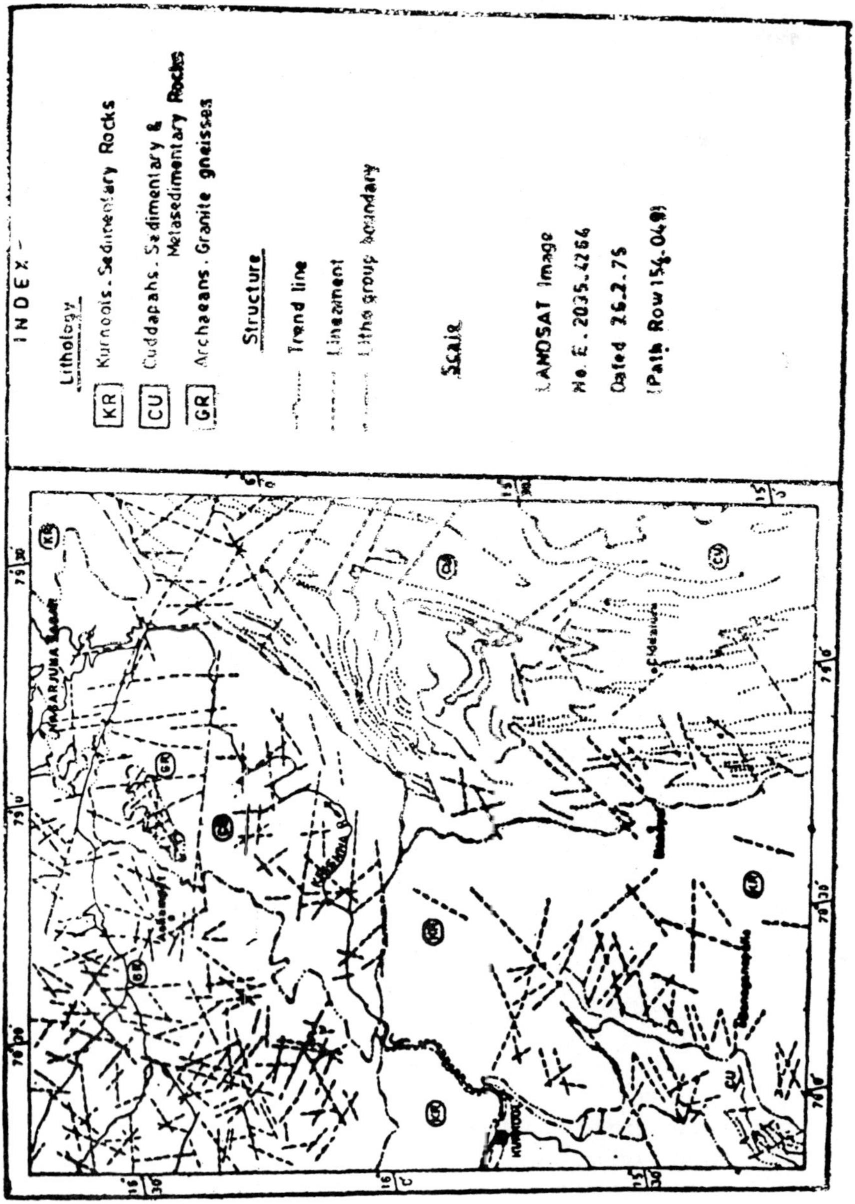

FIG. 16.1 : *General Geology and Structure*

Cuddapahs comprising sedimentary and metasedimentary rocks occupy the area south of Kurnool, form the Srisailam plateau and also constitute the Nallamalai ranges. In Cuddapah area they have swerving strike with gentle to moderate easterly dips. Nallamalai have been twice folded. They have megalineaments in NW-SE to WNW-ESE and in NE-SW to ENE-WSW directions. Kurnools which are sedimentary rocks overlie the Cuddapahs and occupy the area between Kurnool and Banaganapally and north-eastern part of the Srisailam plateau. They have swerving strike and easterly dips. River valleys and intermontane valleys have colluvio-fluvial and fluvial unconsolidated material. There are small patches of blown sand also.

Geomorphology and Land Cover Association

The area can be divided into four 2nd order hydromorphic regional units (Fig. 16.2). In each of them 3rd order erosional as well as depositional macro landforms have been identified. They are described as below :

Pediplain

This occupies the area in north-west and west which is constituted by granitic-gneisses Predominant landform is the coalescing pediments in different stages like active or fresh, dissected or under modification and covered or buried. A prominent piedoment bazada zone is found bordering Srisailam plateau in the east. Streams are of ephemeral nature with narrow flood plains. Several tanks have been constructed across them. Red soils—stoney to clayey, cover whole of this unit except the flood plains where sporadic patches of blackish grey, slightly clayey soil are also found. Riparian vegetation and agriculture is found along these narrow stream lines.

Cuesta terrain around Kurnool and Banaganapalli

Main geomorphic units found in this area are cuestas, fragments of bevelled chestas, a moderately dissected narrow undulating terrain, narrow flood plains occasionally flooded, undulating gullied and imperfectly drained black soil area and the piedmont bazada at the base of the Nallamalai ranges in the east.

All of them have their own landcover associations. Cuestas ... vegetation along fractures. Bevel-
are
the
und
rub

and
al it
nea-
this
shna
The
sur-
oped
iffer-
evel-
exhu-
upy-

t the
exhu-
.3).

with
inter-
ds of
olded
ontrol
Hard
rocks (Quartzites) stand as hog back ridges whereas softer ones (schists, phyllites) have been eroded to give rise to two types of valleys—long open and enclosed bolson type. Both types have considerably thick alluvial fill.

High ridges and their slopes have partly evergreen and mixed forest covers. Bazada and fans have scrub. The flat valleys are cultivated and also have riparian vegetation.

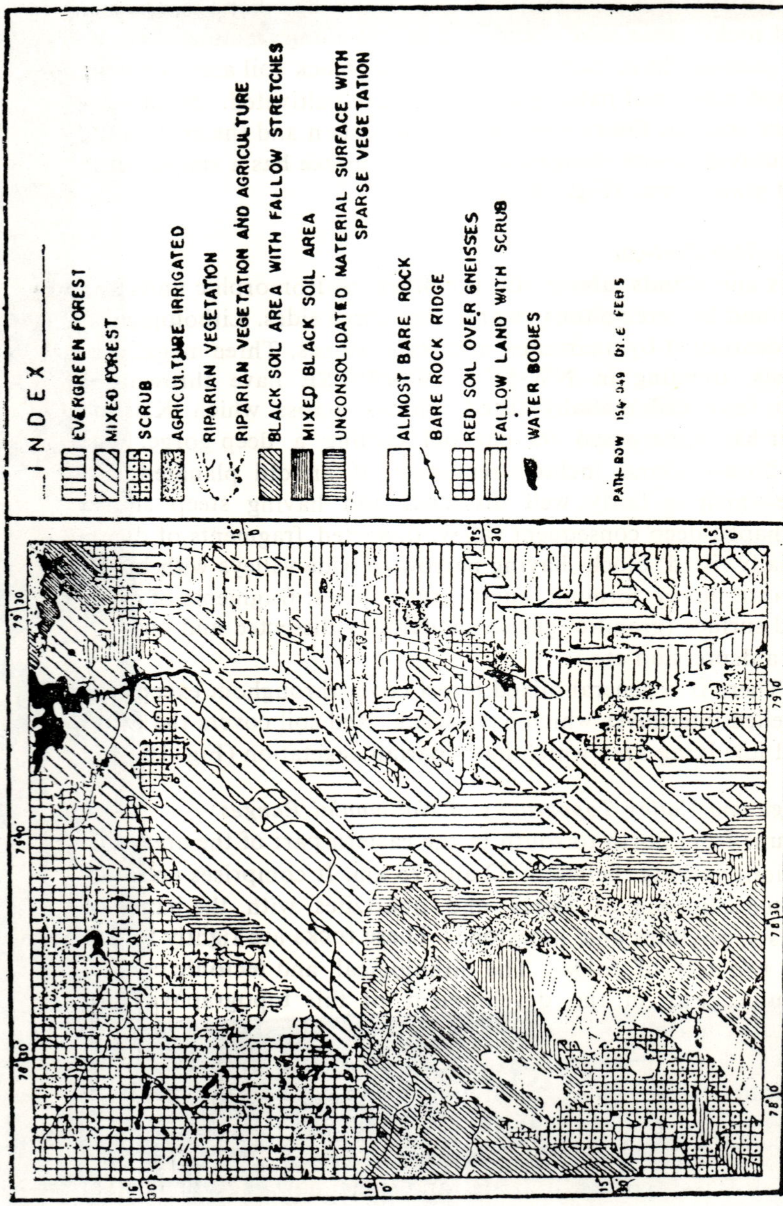

FIG. 16.3: *General Land Cover Categories*

Hydromorphological Analysis

Keeping in view, the above described landform—land cover associations, which also reflect the micro-climatic conditions, the fluvial activity taking place in the area can be analysed. This is done by conceiving a three-dimensional picture about the area indicating percolation, movement and discharge of ground water as influenced by the landform-land cover association.

In the study area, the precipitation water flows along well defined perennial (Krishna river) semi-perennial and ephemeral channels at the top of the Srisailam plateau and along the slopes of Nallamalai ranges. The cuesta slopes also have a few lines of concentrated flow but mostly it experiences a sheet type of flow. While flowing down a part of water percolate sub-surface through fractures in the hard rocks and emerges out at the break of slopes as indicated by the presence of several springs in the Srisailam and Nallamalai hilly area. This reflects the local shallow ground water conditions.

Braided to sheet type of flow takes place on the sparsely vegetated bazada surface. The unconsolidated nature of bazada and fan constituents result a faster sub-surface percolation of water which emerges out at the break of the bazada/fan slope. Therefore, the area around this line bears a very good ground water potential.

In the pediplain the drainage is fracture controlled and the litho base of granitic-gneisses has poor permeability. Sheet wash takes place over the pedimented surfaces which have low density of concentrated flow lines. This results erosion and deposition at random only. The loose sandy material generated due to erosion remains in transit till it reaches the stream channel or flood plain. Limited subsurface percolation takes place in the covered pediments, stream channels and along fractures. Riparian vegetation appearing along these fractures indicates their conduciveness to water percolation.

In the cuesta terrain unit the inter-cuesta depressions and moderately dissected terrain get the slope wash. The infiltration here depends upon the rock type. The undulating black soil area is imperfectly drained indicating impermeable nature of soil cover hence very little water goes sub-surface here. The flood plains receive most of the run-off taking place in this unit. The unconsolidated alluvial constituents make a very good acquifer.

In the Nallamalai ranges high orographic rainfall and thick forest cover result both sub-surface percolation and runoff. Ultimately both of these reach the enclosed narrow filled-in-valleys which form excellent acquifers. Based upon the above analysis potential of each of the landforms has been evaluated. An example of this is given in Table 16.2.

CASE STUDY NO. 2

Desertification studies in a part of Rajasthan (India)

Area Background

The study area forms a part of southern Rajasthan and is bound by latitudes 25° 00′ 00″ and 26° 30′ 00″ N (appx.) and longitudes 72° 00′ 00″ and 73° 30′ 00″ E (appx.). The area shows a transition from semi-arid environs in the south-east to completely arid conditions in the north-west. In its eastern half the area is occupied mainly by the metasedimentary rocks of Ajabgarh group (Delhi Supergroup) and Erinapura and Jalor granites (Post-Delhi) of Proterozoic times. In the western part a few outcrops of sandstone (Jodhpur group) and limestone (Bilara groups) of Marwar Supergroup of Palaeozoic times occur surrounded by sand.

Approach

Desert have a very dynamic environment where changes take place quite often. Sand is one of the most mobile element here. Its generation, mobility and deposition is controlled by the existing landforms, nature of the wind regime and the moisture and vegetation cover conditions. A change in these factors brings in a corresponding change in sand disposition also. Therefore, to find out such changes the behaviour of sand has to be studied from time to time. Such temporal changes now can be easily monitored by studying the Landsat imagery of an area for two or more different periods.

TABLE 16.2 : **Evaluation of ground water potential—Geomorphic province III (Ridge and Valley Topography of Nallamalai Ranges)**

Sl. No.	Land form	Map Anno-tation	Litho Composition	Drainage	Lineaments	Ground water potential	Remarks
1	2	3	4	5	6	7	8
1	Filled in valley floor	AVf	Unconsolidated Colluvio-fluvial material	Braided, sluggish, plenty of water bodies	A few large ones	Very high	Suitable for intensive ground water development, can be taken up directly for geophysical or other sub surface investigation
2	Bolson type filled in valley floor	AVb	Unconsolidated Colluviofluvial material	Centripetal braided sluggish water body in the centre	—	Very high	Suitable for intensive ground water development, can be taken up directly for geophysical or other sub-surface investigation
3	Piedmont Bazada/ fan zone	DSi	Unassorted unconsolidated Colluvio-fluvial material	Parallel, dendritic anastomotic	—	High	Break of the slope zone best suited for development of dug-cum-bore wells

TABLE 16.2—*contd.*

1	2	3	4	5	6	7	8
4	Old now smothered piedmont zone	AWe	Semi to unconsolidated Colluvio-fluvial material	Dendritic & paleo-braiding channels, plenty of water bodies	Moderate density	High	Suitable for dug-cum-bore wells
5	Alluvial plain	AWs	Unconsolidated silt, sand and clayey cover over hard rocks	Imperfectly drained dendritic plenty of water bodies	High	Moderate	Stream vicinities, low areas favourable for dug-cum-bore wells
6	Old Displain remnant	DP	Metasedimentary rocks	Coarse dendritic	Moderate	Low	—
7	Complexes of various types of ridges	RAe, RDe, RHb, RSr, RRt, RUe	Metasedimentary rocks	Trellis, Sub-dendritic angulate, a few water springs	High	Moderate	The valley portions, especially where the water springs are present, suitable for large diameter dug wells

Methodology and Data Inputs
Following data inputs were used for the study:

1. Landsat scene (path-Row No. 160-042) B/W bands 5, 6 and 7 on scale 1 : 1 million, two dates 14th February, 1975 and 3rd October, 1979. (Plate 2, 3)
2. False colour composites of the above.
3. Topographic sheets of the area on scale 1 : 250000 printed in 1968 and 1974.
4. Geological map of Rajasthan prepared by the Geological Survey of India.

The data inputs were visually interpreted. The geomorphic landforms and sand dunes pattern were demarcated on all the three types of data base. These were compared and the changes brought out both qualitative as well as quantitative.

Geomorphology
The area can be divided into three regional morpho-environmental zones: Zone A—the semi arid zone in the southeast, Zone B—The transitional zone in the centre, Zone C—the arid zone in the north-west. Luni river while flowing from north-east to south-west cuts across all of these three zones. (Fig. 16.4 and 16.5).

Zone A
Pali is the main town in this zone. It is characterised by a dissected pediplain where residual hills of granites stand out with irregular outlines in the southern part. They are invariably having a fresh pediment around large hills, like those near Jalor. Fans and bazadas have also developed. At their south-western margins they have semi-established obstacle dunes which are being gullied now.

The pediplain as such has a grey to dark grey coloured soil (paleosole) cover which is now being dissected. There are a few stretches of stone pavements also. Sand has accumulated in some elongated patches. This zone has the maximum agriculture/vegetation cover, amongst all the three zones, concentrated either around tanks or along stream banks.

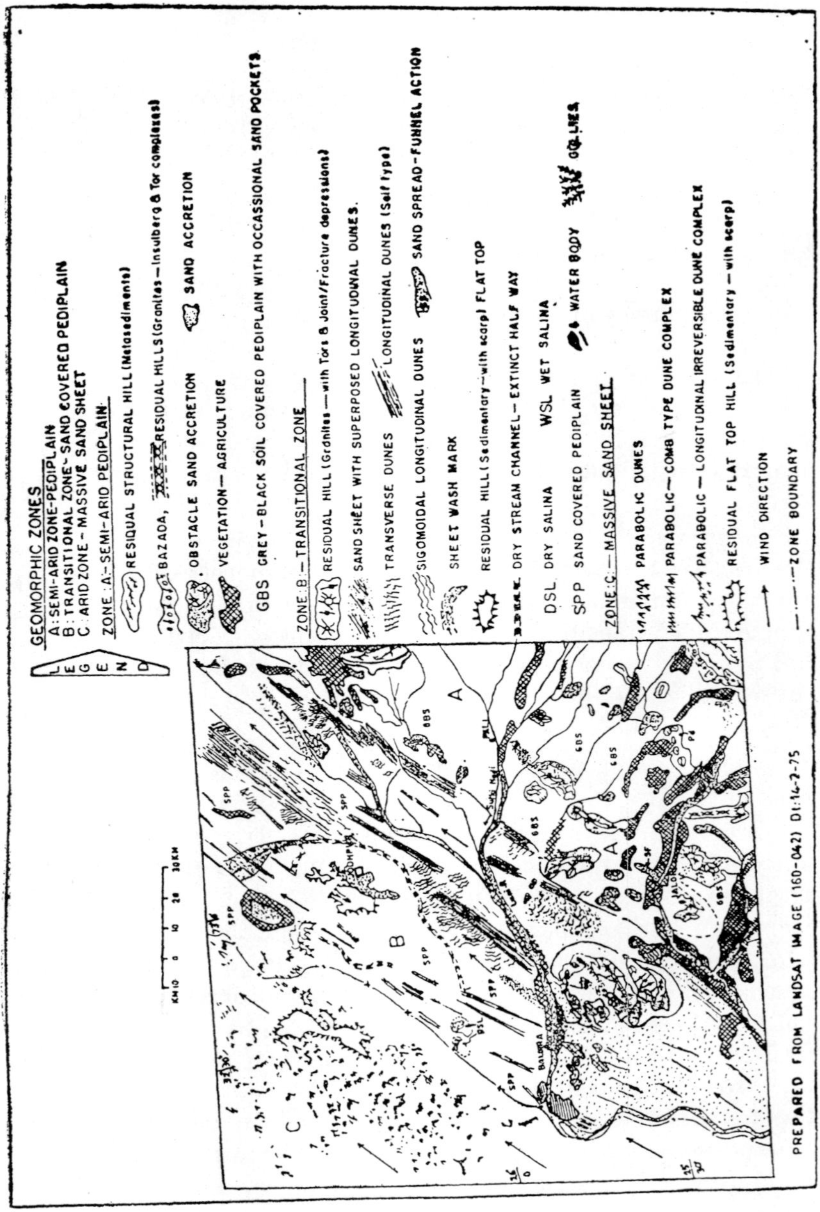

FIG. 16.4 : *Regional Geomorphic Map of Part of Rajasthan*

1. Landsat MSS Image, Band 7, of a part of Andhra Pradesh, India, dt. 26-2-75

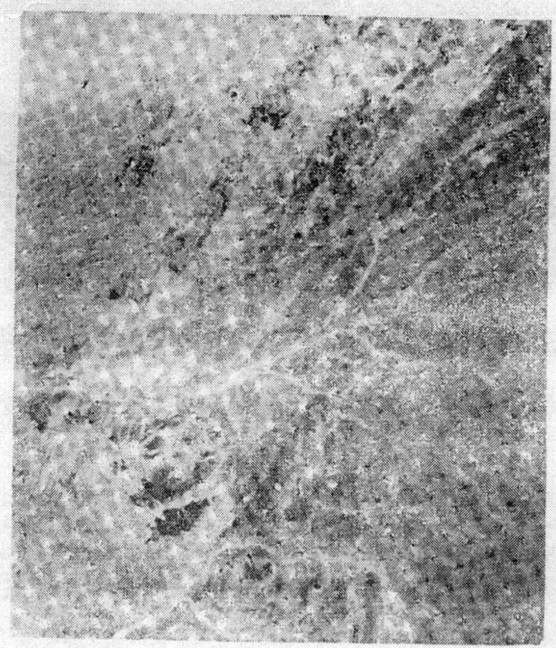

2. Landsat MSS Image, Band 7, of a part of Rajasthan, India Dt. 14-2-75

3. Landsat MSS Image, Band 7, of a part of Rajasthan, India, Dt. 3-10-79

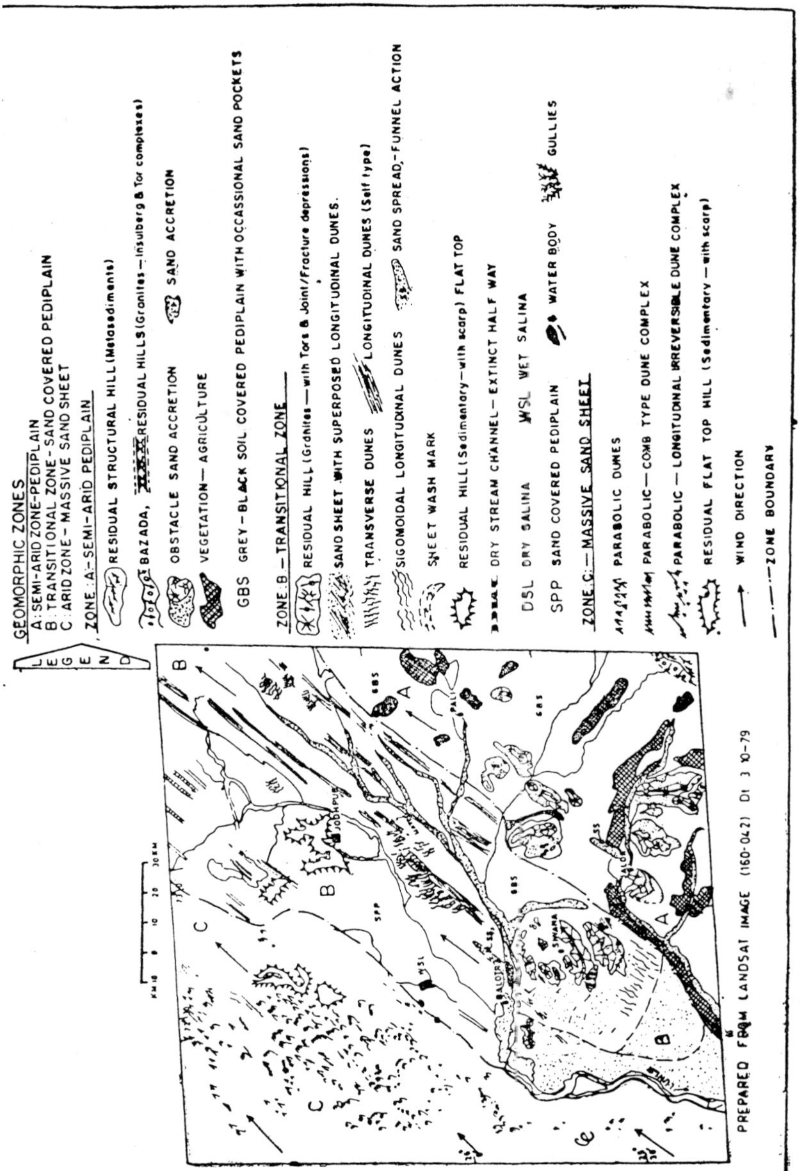

FIG. 16.5 : *Regional Geomorphic Map of Part of Rajasthan*

Zone B

Jodhpur is the main city in this zone which is a transitionary area with mixed conditions. Landforms from both semi-arid and completely arid zones are found. Residual hills of granite stand out in the southern part (around Siwana) forming a concentric pattern. In the north, the sub-horizontal sand stone and lime stone form low scarps. Sand activity is much more than the Zone A and regular dune pattern cover a considerable area. Prominent of them are described as below:

1. South of Luni river and Siwana hills there is an extensive sand sheet with short longitudinal dunes superposed at its surface. Near to the hills to generations of obstacle dunes are formed: (a) The older which is partially stabilised and gullied and (b) the younger which is active. The gaps between two ridges have resulted into funnelling of sand on the northern side.
2. North of Luni river, the sand activity is more intensive. East of Jodhpur, there are two generations of dunes in a NE-SW orientation.
   (a) The earlier transverse dunes and (b) the later longitudinal dunes superposed upon the earlier transverse dunes.
3. The lenticular and sigmoidal sand accretions is NE-SW orientation occurring in the northern part.

Near the western margin of this zone, there is a low lying area with salinas. Agriculture is being practised around Luni river and tanks. Most important feature of this zone is the sand activity which is engulfing more and more area to the east.

Zone C

This is basically a massive sand sheet with parabolic to comb type dune, complexes. In the north-eastern part there is an isolated patch of residual sedimentary rocks standing out with a scarp on all sides. There are areas of more sand activity where the parabolic-comb type of dunes are being disturbed by the blowouts. At places the irreversible transformation of parabolic to longitudinal dunes is also taking place.

Temporal Changes

When the Landsat imagery of two dates viz. 24th February 1975 and 3rd October, 1979 and the older toposheets (1968, 1974) were compared with respect to the above described geomorphic features, it was found that during the period of 11 years (1968 to 1979) all the zones have undergone changes. (Fig. 16.6).

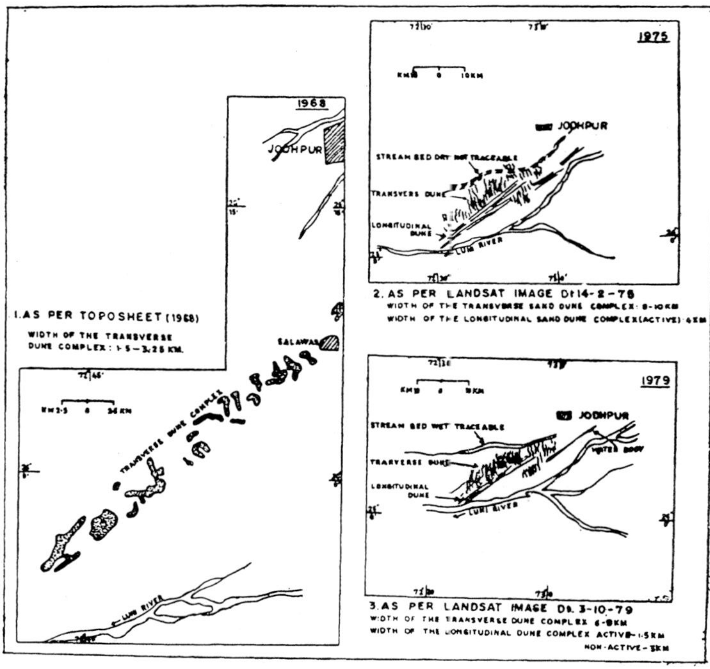

FIG. 16.6 : *Temporal Changes in the Transverse and Longitudinal Sand Dune Complexes—S.W. of Jodhpur*

Important changes found are given below:

1. In the Zones A and B only 5 water bodies are seen in the Landsat image of February 1975. Other water

bodies are dry. When seen in the image October 1979. All these water bodies have larger water cover than before and their number also has swelled to 11.

2. In the Zone B, the course of Jojri Nadi, south-west of Jodhpur, is shown becoming extinct in the toposheet of 1968. This also is seen like that in the February 1975 Landsat image. The picture, however, has quite changed in October 1979. This image shows that the stream had a high amount of water flow and it has extended its channel upto Luni river meeting the master stream west of Balotara.
3. In western part of Zone B the salinas appear dry in the toposheet as well as in the February 1975 image. Later image of October 1979 shows them to be full of water.
4. In the Zone C the parabolic-comb type of dunes have shifted their positions. At the same time, the degree of transformation from parabolic dune to longitudinal dunes has also reduced.
5. The most striking changes in the sand activity seem to have taken place in the Zone B between Jodhpur and Balotra. These changes have taken place as given below:
   (a) Toposheets of 1968 and 1974 show a 3.25 km. wide transverse dune complex to the south-west of Salawas.
   (b) Width of this dune complex has increased to 8—10 km. and a fresh 4 km. wide longitudinal dune complex has superposed on that, as seen in the Landsat image of February 1975. Already existing longitudinal dunes in the north-eastern part have become more prominent. Sigmoidal dunes have been straightened to longitudinal ones.
   (c) It is found in the Landsat image of October 1979 that the width of the transverse dune complex has again been reduced to 6—8 km. The activity of longitudinal dunes has also been reduced as their zone width has come down to 1.5 km. In the northern part also the longitudinal dunes have become less conspicuous and transverse dunes have again reappeared.

All the above mentioned changes in the geomorphic forms indicate the following environmental changes which have taken place in this area.

1. Up to 1968 the moderate dry aridity and wind conditions prevailed in the area. The aridity however started increasing from 1968 onwards as transverse dune complex increased its width. By 1975, the wind velocities also had increased so much that the transverse dune complex surface was stirred and a prominent longitudinal dune zone was superposed on this. At the same time, in the western areas, the change of parabolic dunes to irreversible longitudinal dunes also indicate significant changes of wind velocity—from moderate to fast. In conclusion it can be said that between 1968 and 1975 the area experienced a deflational phase when spread of desertic conditions took place.
2. From 1975 to 1979 the area seems to have undergone stabilising phase. Higher precipitation during 1979 and also possibly during preceding years, brought in stabilisation of sand and growth of vegetation. There was also a reduction in the wind velocities as the longitudinal dune activity was lessened considerably. Moderate winds once again are operative as the transverse dunes have again become prominent.

CONCLUSION

The above two examples from as varied fields as ground water exploration (sub-surface aspects) and the desertification studies (mostly surface aspects) prove that applied geomorphic studies can be fruitfully carried out using Landsat data. The versatility of the data makes it possible to use it in a large number of fields. The studies can be carried out involving less time and cost.

ACKNOWLEDGEMENTS

The author is grateful to Wg. Cdr. K.R. RAO, Director National Remote Sensing Agency to permit to publish this paper and conduct this study.

SUGGESTED READINGS

Bedi, Naresh, 1978: Chapter on geomorphology in satellite remote sensing survey of natural resources of Andhra Pradesh (*NRSA Project Report*.

Bedi, Naresh, 1979a: Regional hydrological studies for groundwater prospecting using Landsat data in two different parts of Southern India. (*NRSA Technical Report No*. 0728).

Bedi, Naresh, 1979b: Larger perspective for geomorphic studies on Landsat imagery. *Proc. 13th Int. Symp. on Remote Sensing of Environment*, ERIM, Ann Arbour, Michigan, USA. (NRSA Technical Report No. 0745)

Air bridge, Rhodes W., 1967: *Encyclopaedia of Geomorphology*, Reinhold Publ. Corp. New Delhi.

Rao, K.R., 1977: Remote sensing for natural resources Key note address at the State convention on natural resources A.P. Academy of Sciences, Hyderabed. (*NRSA Technical Report No*. 0731)

Rao, K.R., 1978: Remote Sensing in Arid Zone Surveys Ins. *Symp. on Arid Zone Research and Development* (NRSA Technical Report No. 0731).

Taranik, James V. & Trantwein, Charles M, 1976: Integration of Geological Remote Sensing Techniques in Sub-surface analysis. USGS. *Open file report No*. 76—402.

Taranik, James V., 1978: Characteristics of the Landsat Multispectral Data System—*Open file report* 78—187, USGS.

Verstappen, H. Th. 1970: Aeolian geomorphology in the Thar desert and paleoclimates. *Zttschr. F. Geom. Sup. Bd*. 10, pp. 104—120.

Verstappen, H. Th., 1972: On dune types, familites and sequences in areas of unidirectional winds. *Gottinger Geogr. Abhand*., Vol., 60 pp. 341—353.

Verstappen H. Th., 1977: *Remote Sensing in Geomorphology*, Elsevier Scientific Publ. Co. Amsterdam.

SEVENTEEN

Sudershan Mukerjee

REMOTE SENSING IN GEOMORPHOLOGICAL MAPPING

THE TERM remote sensing was introduced in the field of geography in 1961, when geography branch of the office of Naval Research, U.S.A., in collaboration with the National Academy of Science Committee on geography contributed a paper on 'Remote Sensing of Environment'. The dictionarical meaning of remote sensing is sensing a thing from a distance which covers even the determination of characteristics of objects through smell, hearing or sight. However in modern science the term remote sensing is confined to the equipments that observe, measure and record energy reflected or emitted in the form of electro-magnetic radiations ranging from ultra-violet to micro wave region.

To differentiate individual objects in some character or other form among the group through remote sensing, the sensors in use are conventional focal length camera, multi-spectral camera, thermal scanner, radiometer, radar, spectro-meter, gravimeter, magnetometer, audiometer etc.

From among the sensors in use to acquire data from aero-space, camera is the prime. With option of lens and the films like panchromatic, colour, infra-red, and colour infra-red in combination with different filters, much can be accomplished with camera. The development of multi-lens camera where more than one lens and now nine lenses with synchronized shutter systems each equipped with suitable film/filter combi-nation so that light of desired bandwidth is allowed to pass through is now in use.

In photographic unit of space imagery the colour infra-red film is in wide use which is useful for acquiring imageries of reflected solar radiation. These imageries are useful in delineation of streams, banks and shorelines accurately. Even the tectonic features are easily detectable on these orbital photographs.

The thermal scanner system is next to camera in obtaining the geographical data through remote sensing. This non-imaging system obtains data through infra-red or multispectral line scanner. The former detects the thermal emission of infra-red region only while the latter is useful for recording data in other portions of electromagnetic spectrum in addition to infra-red. As these scanners have been evolved to detect the thermal or heat energy emitted by objects they are most useful in detecting heated effluence from different objects. From the geomorphological point of view they are the most useful in locating subterranean lava channels that are not visible on the surface and in delineation of water bodies.

The third important sensor is use to sense the geomorphology of the earth is the radar. In case of radar artificial illumination by means of radiation of specific wavelength is done of a target and part of this artificial illumination is reflected back from this target to the detector depending upon the condition of the object irrespective of weather and time of the day. Surface features like faults, scarps and subsurface features like contour beneath the ice cap can be determined accurately by radar system.

In most of the developing countries a good deal of information on geomorphology is still being acquired through ground surveys. Even in developed countries the aerial photography though was in widespread use during 1925-45 yet its practical applicability was superficial. It was during 1955-60 great emphasis was placed on the potentiality of aerial photography and it got its due recognition in the field of geomorphology. It is evident that geomorphology having the study of landforms (including their genesis and development) as its aim has been benefited immensely from aerospace technology and it stands to reason that the recent rapid development of this technology will also contribute to the furthering of geomorphological knowledge since it provides important additional information. The ever increasing quality of the images and their diversity in type, scale etc., enable the expert to map, measure and study a wealth of geomorphological information with unprecedented speed and accuracy. No other means of observation is equal to aerospace imagery in procuring of the geomorphological data. The interpretation of imageries for geomorphological mapping in the first instance requires detection and subsequent recognition and identification of landforms which are afterwards analysed and mapped. In a broader sense it is related to all the geomorphological features that can be observed in the image. The recognition, identification and subsequent analysis of the imagery ultimately produces a map on which apart from individual geomorphological features, larger geomorphological divisions can also be indicated. The stereoscopic study of vertical aerial photographs now can be extended with some further complexity to oblique photographs and it has been introduced already in the interpretation of radar images and orbital photographs too. However the small scale of the orbital photographs limits the possibility of interpretation to major relief features only.

The vertical aerial photographs are the foremost in morphometric analysis of landforms which ranks second in importance in the interpretation and mapping of visible geomorphic features. The relative height of terraces, cliffs etc., the relief amplitude in the area of study, the steepness of slopes and geological dips, the aerial distribution of slope steepness classes, the drainage density, and many other parameters can be measured on vertical aerial photographs with great accuracy.

A third application of aerospace imagery in geomorphological interpretation is to map temporal change. Sedimentation and erosion by rivers, abrasion and accretion of coast lines, changes in the position of glacier tongues, displacement of active dunes and the study of accelerated erosion process, are examples of morphodynamic phenomena that can be successfully ascertained. Best results can be expected from sequential imagery taken at an interval suitable for recording the progressive changes in the landforms concerned. The sequential aerial photographic coverage for the study of geomorphological changes along the lowland coast done from 1967 to 1970 in the SW of Netherlands showed the development in the spit.

The geomorphological study of landforms revealed by receding ice front can also be considerably substantiated by photogrammetric mapping using photographs taken at intervals of several years or decades. The various glacio-morphological phenomena and in particular the diversified accumulation features produced in front of the glacier can also be mapped.

The extensive exercises based on the first Earth Resources Technology Satellite (ERTS-1 renamed since as Landsat-1) in 1972-74 and Skylab-A in 1973 have only confirmed our already widely held belief that remote sensing has much to offer to geomorphology. The Landsat system has been repeatedly recording pictures of the major land areas of the earth ever since it was put into orbit. The information thus procured is now easily available to the geomorphologists for study and mapping the geomorphological features and phenomena.

The resolution of the Landsat system is designed to identify land features having a width of at least 80 metres on the ground. Sometimes sufficiently contrasting linear features hardly 15 metres wide may also come up on the satellite imageries. Elongated features like stream and canal come up very well in one or the other spectral band of these imageries. Often geological features like fault, joint, dyke show up clearly on the satellite imageries.

While photographic reproduction of Landsat imageries is useful for analysis and extraction of information by visual photo interpretation techniques, much more detail information can be obtained by analysis of the basic magnetic tapes in the appropriate computers. It is usually difficult to distinguish visually more than 12 shades while using compatible tapes, it is possible to identify correctly areas having different characteristic of reflection which is more gainful for geomorphological mapping. The repetitive coverage of same area in satellite imageries give an additional advantage to map the changing geomorphological phenomena like courses of rivers in flood affected regions and deltas, extent of snow cover in remote mountains ranges like Himalayas, etc.

India is not now lagging far behind in satellite technology. It launched its first satellite Aryabhata on 19th April, 1975 without any remote sensor on board. However the second satellite Bhaskara launched on 6th June, 1979 carried sensors operating in the visible, near infra-red and microwave frequency bands. The on-board sensors consisted of two TV Camera systems and a microwave radiometer system. The TV camera picked up reflected solar radiation from the ground during day time as Bhaskara passed over India. Though Bhaskara Camera swept almost the whole of India, but relayed the pictures taken over the states of Bihar, Gujarat, Karnataka, Kerala, Madhya Pradesh, Tamil Nadu and Uttar Pradesh only. Rohini-1, the third satellite put into orbit on 18th July, 1980 is carrying multispectral camera and scanner. The ground receiving stations at Sriharikota, Car Nicobar, Ahmedabad and Thumba are receiving the information sent by Rohini through telemetry system. The analysis of the data received from Rohini may add some invaluable detail to the geomorphological map of India.

CONCLUSION

The benefit of application of remote sensing in geomorphological mapping through airborne platform has already been established. The advancement in aerospace technology through launching of statellite fitted with sensors like camera and scanner has given a fresh impetus for doing the geomorphological mapping through remote sensing. The advantages of survey from orbital altitude includes synoptic overview, the potential for frequent coverage and a data collection capacity in areas otherwise inaccessible or under hostile conditions. Earth orbiting satellites are presently capable of providing imagery from a wide array of devices—conventional and T.V. cameras as well as sensing instruments which detect difference in heat radiations and light reflectance from the surface below. Assuring continued successful development and use of these instruments, remote sensing from space platform will return vast quantities of information on and to the earth particularly for geomorphological mapping.

It is hard to think of a more congruous discipline than geomorphology with regard to profitable use of data from space. Thus many applications will remain potential rather than practical unless more geomorphic research is developed. Geomorphology needs experienced investigators using these new tools along with geologists, hydrologists, meteorologists, oceanographers, physicist and engineers to expand research into areas that cannot be penetrated with traditional techniques and instruments and to map our planet in a better way.

SUGGESTED READINGS

Barrett, B.C. and Curtis, L.F., 1976: *Introduction to environmental remote sensing*, Chopman and Hall, London.

Bird, J.B.; Morison A., 1964: Space photography and its Geographical Applications, *Geographical Review*, V. 54, pp. 464-486.

Estates, J.E., 1966: Some Applications of Aerial Infrared Imagery, *Annals of the Association of American Geographers*, V. 56, pp 672-682.

Lancaster, Jane, 1968: Geographers and Remote Sensing, *The Journal of Geography*, Vol. 68, pp. 301-310.

M. Lawrin, John D., 1971: Application of Remote Sensing to Planimetric, Thematic and Topographical Mapping, *Proceedings of the International Workshop of Earth Resources Systems* V. II office of space science applications, Washington, pp. 171-194.

Meijerkink, A.M.J., 1976: Digital Pre-processing Techniques of Landsat Data in High Relief Area—A Himalayan Example, *Photonirvachak*, Vol. 4, No. 1 & 2, pp. 15-19.

Mukerjee, Sudershan, 1972: Remote Sensing in the field of Geography, *Geographical Review of India*, Vol. XXXIV, No. 2, pp. 178-183.

Mukerjee, Sudershan, 1977: Geographic Data from Space, *The Indian Geographical Journal*, Vol. 52, No. 1 & 2, pp. 32-33.

Newhall, Beaumont, 1961: *Airborne Camera—The world from the Air and Outer Space*, Hastings House Publishers, New York.

Rao, D,P. 1977: Visual comparison of computer aided landsat map and airphoto map. A case study from Karimnagar district, Andhra Pradesh, *Photonirvachak*, Vol. 5, No. 1, pp. 25-30.

Schmitt-Taverna, M.R., 1975: The West plaat (SW Netherlands)—A study of Coastal Dynamics from Sequential aerial photography, *ITC Journal* 3(2), 175-185.

Sharma, R.P. 1974: The Application of Hyper-Altitude photographers for Geological Mapping, *Photonirvachak*, V. 2, No. 1, pp. 1 -11.

Simpson, R.B., 1966: Radar, Geographic Tool, *Annals of the Association of American Geographers*, Vol. 56, pp. 80-96.

Singh, B.N., 1977: Interpretation of Satellite Imagery for Delineation of Ravines, *Photonirvachak*, Vol. 5, No. 1, pp. 31-34.

Srinivasan, R. and Srinivasa, B.L., 1977: Some new Geological Features from the Landsat Imagery of Karnataka, *Journal of Geological Society of India*, Vol. 18, No. 11, pp. 589-597.

Subbarao, B.V.; Alasingrachar, M.A.; and Shah, S.D., 1980: Application of Aerial photographs and Landsat Imagery for Geological Mapping around Chandrapur (Maharashtra), *Journal of the Geological Society of India*, Vol. 21, No. 2, pp. 84-90.

Verstappen, H. Th., 1977: *Remote sensing in geography*, Elsevier Scientific Publishing Company, Amsterdam.

EIGHTEEN

Surendra Singh

ROLE OF PHOTOGEOMORPHOLOGICAL TECHNIQUES IN THE INTEGRATED NATURAL RESOURCES SURVEY OF THE CENTRAL LUNI BASIN, WESTERN RAJASTHAN (INDIA)

INTRODUCTION

IN THE ARID and semi-arid regions of India, the human and livestock population is increasing at a faster rate. The production per unit of land holding is low and inadequate which is not keeping pace with the increase in population. To cater the basic food needs and to improve the living conditions of the people, a full appraisal of the natural resources, their geographical distribution, limitations, present use and scope for better use is essential. To achieve this aim, the efficient methods and techniques of survey and mapping are essential pre-requisites. Now-a-days, the aerial photographs are the most important tool in preparing the inventory of all the natural resources. The natural resources except the landforms are not directly seen on the aerial photographs under the stereoscope. The landforms being the most conspicuous features are directly visible in three-dimensions on the aerial photographs, the knowledge of photogeomorphology for the reliable photointerpretation of the other natural resources is of paramount importance. The scientists of these disciplines, therefore, have to base their conclusions on 'indirect' evidence which is mostly of geomorphological nature.

The significance of photogeomorphological techniques in the natural integrated surveys have been discussed by Verstappen (1967). The importance of geomorphology in evaluating the individual resource has also been described by several earth scientists. Tator (1958) emphasised the use of geomorphic criteria in the solution of 'hidden geology' problems. Goosen (1964) has stated that the most of the soil surveyors agree that the grouping of soil units on a map must be done on a geomorphological or landscape basis, irrespective of the kind of taxonomic units used in defining the mapping units. Mallick and Niyogi (1972) have discussed the application of geomorphology in ground water prospecting in the alluvial plains around Burdwan, West Bengal. Saxena (1972) has described the vegetation types of Western Rajasthan in relation to landforms.

In the present article, the role of photogeomorphological techniques in the integrated natural resources survey of the Central Luni basin situated between lat. 25° 0′ to 26° 20′ N and long. 71° 50′ to 73° 30′ E and covering an area of 16528.0 sq. km. has been discussed.

METHODS AND MATERIAL

Prior to the field work, the aerial photographs of 1 : 40000 and 1 : 31000 scales were geomorphologically interpreted, different landforms were identified and their boundaries were delineated. After geomorphological photointerpretation, the same aerial photographs were interpreted by the other earth scientists and they demarcated the boundaries of soil, vegetation and water resources in each landform unit already delineated by the geomorphologist. Field traverses and query points were selected in different landforms and the field traverses were undertaken by the inter-disciplinary team of scientists to check and to correct the variations perceived in the boundaries of the resources from the aerial photo-interpretation. Some additional information on different natural resources which could not be collected from the aerial photographs were collected in the field from the each landform along the selected query points. After field survey, the aerial photographs were again interpreted, and data collected from field were incorporated. Each scientist prepared the individual resource map by using landform map of 1 : 253440 scale as a base map. Finally, these resource maps were superimposed on each other and a map of 1 : 506880 scale (Fig. *18.1*) showing the Major Land

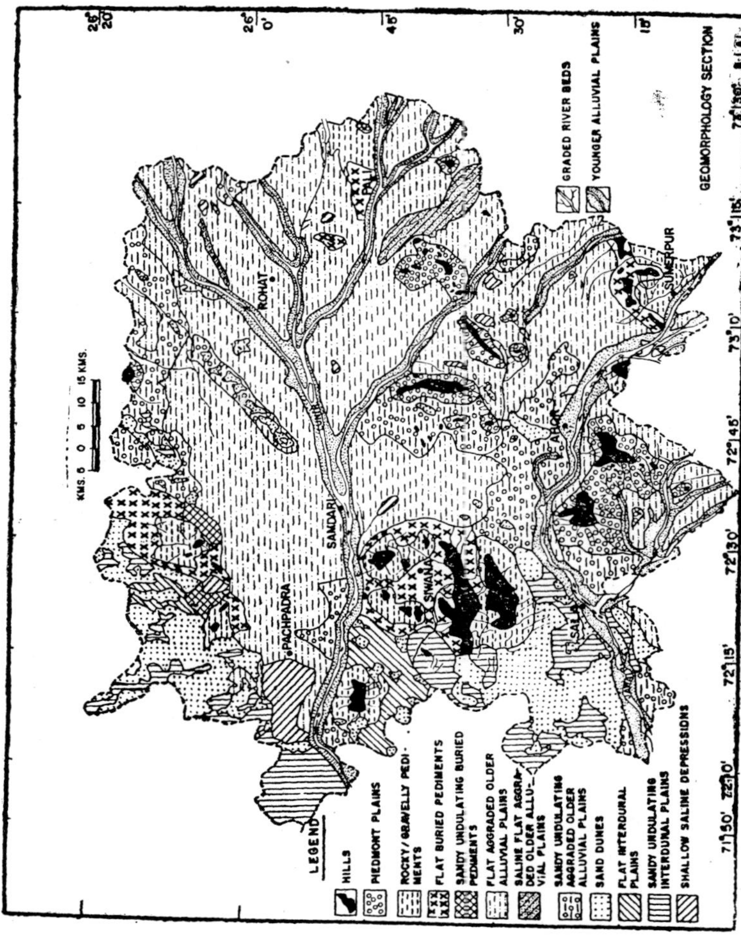

FIG. 18.1 : *Major Land Resources Units (Landform Units) of the Central Luni Basin*

Resources Units (MLRU) was prepared by using the photogrammetric method described by Ghose and Singh (1966).

RESULTS AND DISCUSSION

By employing the photogeomorphological techniques, the basin has been divided into the following 14 landform units which have been termed as Major Land Resources Units and their resources have been integrated at one place. The salient characteristics of these landforms or resource units are presented in Table 18.1. The impact of the geomorphology on the distribution of the natural resources of the basin is discussed below.

Impact of Geomorphology on Natural Resources

Hills (Area—575.44 sq. km.)
The hills are located scattered and they constitute the phyllite, schist, granite, rhyolite, limestone and sandstone rocks. The slopes of these hills vary from 20° to 35° and are covered with debris of 3 to 60 cm diameter. The radial and subparallel drainage patterns are developed along the joints.

This landform is devoid of soils except shallow skeletal soils occur along the joints. The dominant plant communities of this landform are *Anogeissus pendula-Acacia senegal*, *Acacia senegal-Euphorbia caducifolia.* The associates are *Grewia tenax* and *Commiphora wightii*. Limestone formations are good host of ground water and it is found at 25 to 40 m depth. Surface run-off has been collected in tanks and reservoirs along the joints. It is a rocky waste and is not suitable for cultivation.

Piedmont Plains (Area 889.05 sq. km.)
The plains are located at the foot of the hills and are composed of heterogeneous sediments. The slopes vary from 2° to 5°. Dendritic and sub-parallel drainage patterns are formed. This unit is affected by severe to very severe water and wind erosion. The soils associated with this landform are deep to very deep sand to loam sands underlain by unconsolidated rock fragments. This landform is dominated by *Euphorbia caducifolia-Acacia senegal* community with 60% and 30% relative dominance. The associates are *Maytenus emarginata* and *Acacia*

TABLE 18.1 : **Salient characteristics of the natural resources of Major Land Resources Units (Landform Units)**

S. No	Major Land Resources Units (Landforms)	Geomorphology	Soils	Vegetation	Water resources	Land use
1	2	3	4	5	6	7
1.	Hills	Rhyolite, granite, sandstone, limestone, quartzite and schist are the chief constituents. Slope 20° to 35°; dendritic and radial patterns and severe to very severe water erosion	Devoid of soils except shallow skeletal soils in pockets	*Anogeissus pendula*, *Acacia senegal*, *Grewia tenax* and *Euphorbia caducifolia*	Good reserves of ground water in limestone cavities. Surface water storage in the tanks along the joints	Rocky waste, not suitable for cultivation.
2.	Piedmont plains	Situated at the foot of the hills, slope 2° to 5°,	Deep to very deep sands to loamy sand	*Euphorbia caducifolia*, *Acacia senegal*	Ground water is not available. Surface	Single cropped, bajra and guar occa-

(*contd.*)

TABLE 18.1.—*contd.*

1	2	3	4	5	6	7
		sub-parallel and dendritic drainage patterns and severe to very severe water and moderate wind erosion	soils underlain by rock fragments	and *Maytenus emarginata*	water is stored in tanks	sionally grown
3.	Rocky/Gravelly pediments	Rocky surface; slope 0° to 2°, dendritic drainage pattern and moderate to severe water erosion	Devoid of soils, except shallow heterogeneous soils in pockets	*Maytenus emarginata*, *Capparis decidua* and *Acacia senegal*	Ground water encountered at 12 to 20 m depth and surface waters collected in tanks	Rocky-gravelly waste and is not suitable for cultivation
4.	Flat buried pedimonts	Almost level with 0° to 1° slope, sub-parallel drainage pattern, moderate water erosion only among the channels	Moderately deep to very deep loamy sands to sandy loam and loam soils underlain by weathered rocks	*Prosopis cineraria*, *Acacia nilotica* and *Acacia senegal*	Ground water is found at 7 to 10 m depth	Partly single cropped and partly double cropped, with bajra, moong, wheat and chilies

5. Sandy undulating buried pediments	Sandy deposits in the form of hummocks and dunes; slope 1° to 3°, severe to very severe wind and water erosion	Very deep sandy to loamy sand soils underlain by hard rock	*Acacia senegal*, *Calligonum polygonoides* and *Calotropis procera*	Ground water at certain places occurs at 20 to 30 m depth	Single cropped, occasionally cultivated with bajra, guar, and moong
6. Flat aggraded older alluvial plains	Flat land with 0° to 1° slope; sediments underlain by kankar pan at 7 to 180 cm depth, slight to moderate water and wind erosion	Shallow to moderately deep and deep to very deep loamy sands to sandy loam and also loam soils underlain by lime concretionary zone	*Prosopis cineraria*, *Zizyphus nummularia* and *Capparis decidua*	Ground water along the prior drainage channels found at 8 to 15 m depth. Surface water stored in numerous tanks	Every year cropped with bajra, guar, moong and moth. Double cropped in pockets
7. Saline flat aggraded older alluvial plains	Level land with 0° to 1° slope, sheet flow dominant, moderate to severe salinity hazards	Medium to heavy textured loam to silty clay loam and also clay loam saline soils underlain by segregated amorphous lime	*Salvadora oleoides*, *Acacia nilotica*, spp. *indica*	Saline ground water along prior drainage channels at 10 to 15 m depth	Kharchia wheat in rabi and bajra in kharif are cropped

(contd.)

TABLE 18.1—*contd.*

1	2	3	4	5	6	7
8.	Sandy undulating aggraded older alluvial plains	Sandy deposits in the form of hummocks, sand sheets and low dunes lies on alluvial deposits and *kankar* pan. Slope 0° to 3° and moderate to severe and very severe wind erosion	Deep to very deep sands to loamy sands and sandy loam soils underlain by lime concretionary zone	*Calligonum polygonoides*, *Prosopis cineraria* and *Zizyphus nummularia*	Ground water only along the prior drainage channels found at 18 to 30 m depth	Single cropped, occasionally cultivated with bajra, guar and moong and wheat is raised in pockets under irrigation
9.	Sand dunes	Stabilised dunes such as longitudinal, transverse, parabolic and obstacle and active barchan and shrub-coppice dunes. Stabilised dunes have lime nodules and 2° to 20° slope. Very severe wind and severe water erosion	Very deep uniformly layered calcareous sands with clay content 4 to 5% and silt more than 1%	*Acacia senegal*, *Calligonum polygonoides*, *Clerodendrum phlomoides* and *Prosopis cineraria* on stabilised dunes. No vegetation on active dunes	Ground water is not available, surface water collected in tanks at the junction of coalesced dunes	Stabilised dunes occasionally cultivated with bajra and guar. Active dunes are not cultivated

10. Flat inter-dunal plains	Almost level with 0° to 1° slope and occur in between the stabilised dunes. *Kankar* pan at varying depth encountered and slight to moderate water and wind erosion	Moderately deep to deep and very deep loamy sand to sandy loam soils underlain by hard *kankar* pan	*Prosopis cineraria, Capparis decidua* and *Zizyphus nummularia*	Ground water at 15 to 20 m depth encountered along the prior drainage channels and surface water stored in numerous tanks	Every year cropped with bajra, guar til and moong. Wheat is grown in pockets
11. Sandy undulating interdunal plains	Sand sheets, low dunes and hummocks cover the alluvial surface; slope 0° to 3°, moderate to severe and very severe wind erosion	Deep to very deep calcareous sands to loamy sands and also sandy loam underlain by lime concretionary strata	*Calligonum polygonoides, Capparis decidua* and *Zizyphus nummularia*	Both ground and surface water resources are not good. Ground water found in pockets only at 8 to 15 m depth	Single cropped, occasionally cultivated with bajra, guar and mung
12. Shallow saline depressions	Shallow depressions of different shape and size occur at river confluences; sheet flow dominant; slope 0°to 1°. Very severe salinity hazard	Deep to very deep highly saline loam to clay loam soils underlain by sands and lime concretionary layer	*Halophytic* species like *Peganum harmala, Suaeda fruticosa & Sporobolus*	Highly saline ground water occurs near the confluences at 10 to 15 m depth	Not suitable for cultivation but is a good gathering ground for natural evaporites

(*contd.*)

TABLE 18.1—*concld.*

1	2	3	4	5	6	7
13.	Graded river beds	Meandering and braided streams, severe bank erosion dominant at places, slope 0° to 1°; moderate wind and salinity hazards at places	Saline and non-saline heterogeneous saline soils with poor water holding capacity	*Tamarix dioica Aerva Persica*, *Acacia jacquemontii* and *Tamarix articulata*	Ground water found in pockets at 10 to 12 m depth, surface water collected in reservoirs	Beds are not cvltivated; sandbars are occasionally cultivated with bajra
14.	Younger alluvial plains	Slope 0° to 1°, subparallel drainage pattern, moderate water and wind erosion	Coarse to medium textured sands to loamy sands & sandy loam soils, absence of *kankar* pan	*Salvadora oleides*, *Prosopis cineraria* & *Acacia nilotica*	Sufficient reserves of groundwater are encountered at 7 to 12 m depth grown	Mostly double cropped with wheat, chillies & vegetables, kharif crops are at some places

pendula. The condition of the ground water due to the presence of hard rock below the sediments is not good. Surface water potentials have been collected in tanks at the lower slopes of this landform. This landform is occasionally cultivated for growing bajra, moong, moth and guar.

Rocky/Gravelly Pediments (Area 788.95 sq. km.)

This landform with 1° to 2° slope and veneer deposits in pocket lies at the foot of the hills of different formations. The joints and cracks are developed on the surface. The dendritic and sub-parallel drainage patterns are formed here. The water action is moderate to severe. This landform is devoid of soils except some gravelly sands to loamy sand soils in pockets. They are dominated by *Euphorbia caducifolia Acacia senegal, Acacia senegal Capparis decidua* and *Acacia senegal Maytenus emarginata* communities. This landform has good potentials of ground and surface waters. The ground water occurs at 12 to 20 m depth. Surface water has been collected in tanks for human and livestock consumption. This landform is not put under cultivation but is used as pasture land.

Flat Buried Pediments (Area 867.26 sq. km.)

This landform is nearly level with 0° to 1° slope. It is covered with alluvial and colluvial sediments of 20 cm to 1 m thickness underlain by hard rock. It is affected by moderate water erosion only along the drainage channels and rest of the area is free from erosion. The soils developed on this landform are brown to dark brown loamy sand to sandy loam underlain by weathered and solid rocks of different formations. This landform is dominated *Prosopis cineraria-Acacia nilotica* and *Euphorbia caducifolia-Acacia senegal* communities. Due to the rapid rate of infiltration in the underlying weathered strata, the ground water potentials are good. The ground water in this landform has been encountered at 7 to 10 m depth. This landform sustains both irrigated and unirrigated crops. Double cropped like wheat, chillies and vegetables are raised under irrigated conditions. The unirrigated crops such as bajra, moong and moth are grown.

Sandy Undulating Buried Pediments (Area 70.05 sq. km.)
The rocky pediments covered with sand sheets, dunes and hummocks of different thickness and height are termed as sandy undulating buried pediments. The slope is irregular varying from 1° to 3°. The prodigious sandy deposits have been dissected by the numerous drainage channels of 5 to 10 m depth and 10 to 20 m width. This unit is affected by severe to very severe wind and water erosion. The soils are sandy, single grained and calcareous. They have poor water holding capacity. The dominant plant community of this landform is *Acacia senegal-Calligonum polygonoides-Calotropis procera*. The scattered plants of *Prosopis cineraria* and *Maytenus emarginata* are also present. The condition of the ground water due to rapid infiltration through sandy deposits is good and water is found at 20 to 30 m depth. This landform is not agriculturally potential and is occasionally cultivated for growing bajra and guar.

Flat Aggraded Older Alluvial Plains (Area 6326.65 sq. km.)
These plains with 0° to 1° slope are quite extensive. The sediments are composed of rounded to subrounded grains of quartz, mica, iron and hornblende. The sediments are underlain by *kankar pan* at 7 to 180 cm depth. This unit is free from erosion excepting at few places where it is affected by slight to moderate water and wind erosion. The soils of this landform are of varying texture such as sandy to loamy sand and loamy sand to loam. The soils are underlain by compact lime concretionary zone. The most prevalent plant community of this landform is *Prosopis cineraria-Capparis decidua-Zizyphus nummularia*. Few associated trees and shrubs are *Acacia nilotica* ssp. *indica*, *Tecomella undulata*, *Lycium barbarum* and *Calotropis procera*. The condition of ground water is only good along the courses of the prior channels and water occurs at 8 to 15 m depth. In other parts of this landform, surface water potentials are good and numerous tanks exist here. The double cropping is practised in those parts where ground water is available and the crops like wheat and chillies are raised. This landform is intensively cultivated for growing kharif crops like bajra, guar, moong and moth.

Saline Flat Aggraded Older Alluvial Plains (Area 171.79 sq. km.)
These plains are also flat with 0° to 1° slope but have become saline due to impeded drainage conditions. The sediments are coarse to medium textured. This unit is free from erosion but moderate to severe salinity hazards have affected the lands of this unit. The soils of this landform are heavy textured loam to silty clay loam and also clay loam. These soils are moderately saline and are underlain by thick deposits of segregated amorphous lime deposits. This landform is dominated by *Salvadora oleoides-Prosopis cineraria-A. nilotica* ssp. *indica.* The saline ground water occurs along the prior drainage channels at 10 to 15 m depth. Due to salinity hazards, the agricultural potentials of this landform are not good. The irrigated crops like Kharchia wheat are grown with saline water.

Sandy Undulating Aggraded Older Alluvial Plains (Area 1649.67 sq. km.)
Light to medium textured alluvial plains were degraded by aeolian activities and they have created undulations in the form of hummocks, and low transverse and longitudinal dunes. The slopes are irregular varying from 0° to 3°. The *kankar pan* has been encountered at 200 cm depth and this unit is affected by moderate to severe and very severe wind erosion. The soils of this landform are deep to very deep sands to loamy sands and also sandy loam. The soils are well drained with rapid permeability. The areas between the hummocks and ridges are dominated by *Prosopis cineraria-Zizyphus nummularia* community. The undulating parts are occupied by *Zizyphus nummularia-Calligonum polygonoides* and *Calligonum polygonoides-Lasiurus sindicus* plant communities. The condition of the water along the prior drainage channels is satisfactory, and water is available at 18 to 30 m depth. This landform is only capable to sustain single kharif crops and is kept fallow for two to three years.

Sand Dunes (Area 1474.54 sq. km.)
The dunes are of two systems viz. dunes of old system and dunes of new system. The parabolic, longitudinal, transverse and obstacle dunes belong to the old dune system. The dunes of this system are stabilised, well vegetated and fossilised. The

obstacle dunes are dissected by the gullies of different dimensions. The height of these dunes vary from 8 to 70 m and the slopes of the windward, leeward sides and flanks are 2°, 2° to 4° and 8° to 12° respectively. The dunes of new system are barchans, hummocks, low longitudinal and transverse ridges. They are of 45 cm to 1 m and 3 to 7 m heights. These dunes are active and they cause lot of damage to the agricultural lands and to the other permanent structures. This landform is affected by severe to very severe wind and water erosion. The soils of the stabilised dunes are very deep, uniformly layered calcareous sands. The clay content ranges from 4 to 6% and silt more than 1%. The sands are hard and compact. The soils of the active dunes are uniformly layered loose and non-calcareous sands. The dunes of old systems are dominated by *Acacia senegal-Calligonum polygonoides, Calligonum polygonoides-Clerodendrum phlomoides. Prosopis cineraria-Calligonum polygonoides, Maytenus emarginata-Salvadora oleoides* and *Acacia jacquemontii-Calotropis procera* plant communities. The active dunes do not support any vegetation. The dunes are not good host of ground water. The surface water potentials are good and they have been collected in tanks at the lower slopes of the coalesced parabolic dunes. The dunes of old system are occasionally cultivated for growing bajra, moong and guar. The active dunes are not cultivated.

Flat Interdunal Plains (Area 258.53 sq. km.)

This plains occur scattered between the dunes of old system in the north and northwestern parts of the region. They are almost level with 0° to 1° slope. They are free from erosion except slight to moderate water and wind erosion in pockets. The sediments are underlain by *kankar pan* at 30 to 170 cm depth. The soils of this landform are moderately deep to very deep loamy sand to sandy loam and also loam underlain by indulated concretionary zone. *Prosopis cineraria-Capparis decidua* and *Zizyphus nommularia Capparis decidua* are the dominant plant communities of this landform. The potentials of ground water in Padru-Dekhan sector is good and water is found at 15 to 20 m depth. This landform has also good potentials of surface water and is dotted with numerous tanks.

It is a single cropped land and the crops like bajra, til, guar, moong and moth are raised here.

Sandy Undulating Interdunal Plains (Area 967.44 sq. km.)
This landform also occurs scatterly in between the stabilised dunes of old system in north and northwest parts of the region. The alluvial surfaces of this unit are covered with sands and low dunes which have created undulations. The slope varies from 0° to 3°. This landform suffers from moderate to severe and very severe wind erosion. These factors have affected the physical potentials of this unit. The soils of this landform are deep to very deep sands to loamy sands. They are highly calcareous and are underlain by weekly developed concretionary zone of carbonates. The dominant plant communities of this landform are *Calligonum polygonoides-Zizyphus nummularia*, *Calligonum polygonoides-Capparis decidua* and *Calligonum polygonoides-Haloxylon salicornicum*. The ground water along the prior drainage channels has been found at 8 to 15 m depth. Due to physical limitations, this landform is occasionally cultivated and is kept fallow for 3 to 4 years.

Shallow Saline Depressions (Area 348.59 sq. km.)
The saline depressions of different shape and size are situated near Pachpadra, Sanwarla, Thob, Dedas and Samuja villages. The surfaces of this unit are sealed and highly saline. It is nearly level with 0° to 1° slope and free from erosion. The sodium chloride, calcium sulphate and calcium carbonate are extracted from these depressions. The soils of this landform are heavy textured silty clay loam to clay loam and are highly saline and sodic, underlain by gypsiferous and calcareous material. The halophytic species like *Peganum harmala*. *Sporobolus helvolus*, *Cressa cretica*, *Suaeda fruticosa* and *Chloris virgata* occur on this landform. Highly saline water near the confluences of the prior drainage channels occur at 10 to 15 m depth. This unit is not suitable for cultivation.

Graded River Beds (Area 522.06 sq. km.)
The beds of the Luni, the Jawai, the Sukri, the Khari, the Bandi, the Mitri and the Golasni rivers are graded with 0° to 1° slope. The meandering and braiding patterns are developed in

river courses. The river beds are covered with sediments of 0.44 to 0.45 mm diameter. The sand bars of 500 to 700 m length; 250 to 300 m width and 1 to 2 m heights are formed in the beds of these rivers. Severe bank erosion is dominant at places. The soils of this unit are gravelly sands and grit and have poor water holding capacity. The soils of the beds at certain places are saline. The soils of the sand bars are hard and compact sands. The sand bars are occupied by *Tamarix dioica. Tamarix articulata, Leptadenia pyrotechnica, Aerva persica* and *Acacia jacquemontii.* The ground water potentials at some places are good and water is found at 10 to 12 m depth. This unit is not fit for cultivation.

Younger Alluvial Plains (Area 1617.98 sq. km.)
This unit occurs on either side of the banks of the rivers. These plains are almost level with 0° to 1° slope. The drainage pattern is sub-parallel. They are free from erosion except at few places where they are affected by moderate wind and water erosion. The sediments are not underlain by *kankar pan.* The soils of this landform are coarse to medium texture varying from sands to loamy sands and also sandy loam to loam. The soils at certain places are saline and have poor permeability. The common plant community of this landform is *Salvadora oleoides* associated with *Prosopis cineraria, Acacia nilotica* and *Capparis decidua-Salvadora oleoides-Capparis decidua* and *Salvadora oleoides-Prosopis cineraria* occur at distinct communities. The ground water potentials of this landform are better than other landforms due to larger density of wells. The ground water has been encountered at 10 to 15 m depth. Due to the availability of sufficient ground water, this landform is intensively cultivated and double cropping is practised. Kharif crops like, bajra, guar, moong and moth are also grown in the areas where ground water is insufficient.

CONCLUSIONS

1. The classification and analysis of landforms by employing geomorphological techniques has played a key role in the integrated natural resources survey of the Central Luni basin.

2. In the basin 14 land resource units (landform units) have been identified and mapped. The resources of each unit have been integrated at one place which revealed that different resource units have different production potentials.
3. The younger alluvial plains, flat buried pediments and some parts of flat aggraded older alluvial plains due to the availability of ground water have the good agricultural potentials for the irrigated crops. Flat aggraded older alluvial and interdunal plains are most suitable resource units for raising rainfed crops like bajra, til, moong, moth and guar.
4. Sandy undulating aggraded older alluvial and interdunal plains, piedmont plains and sand dunes could be used for the development of grasslands. The flat areas of these resource units have good agricultural potentials for growing the rainfed crops.
5. The hills and rocky pediments and parts of flat aggraded older alluvial and interdunal plains underlain by hard *kankar pan* have good surface water potentials and could be developed into desert catchments. The saline depressions are the gathering grounds of natural evaporites.

It may, therefore, be concluded that integrated inventory of the resources, their proper assessment and management will enable the planners to develop the resources of these land resource units for the overall development of the basin. This will ultimately help to upgrade the yield of the water resources and agricultural production to improve the living conditions of the people.

ACKNOWLEDGEMENT

The author wishes to express his sincere thanks to Dr. H.S. Mann, Director, Dr. K.A. Shankarnarayan, Head of Basic Resource Division and to Shri Bimal Ghose for encouragement and guidance to write up this article. Thanks are also due to the staff of Geomorphology section for help in compilation of survey data.

SUGGESTED READINGS

Goosen, D., 1964: The use of aerial photography for the development of soil resources in South America. *Archives of Photogramm.* XV (Lisbon I.S.P. Congress).

Ghose, B. and Singh, S., 1966: Quick method of aerial photogrammetric mapping of geomorphic features, soil and vegetation community. *Ind. J. Geog.* 1(1), 49-56.

Mallick, S. and Niyogi, D., 1972: Application of Geomorphology in ground water prospecting in the alluvial plains around Burdwan, West Bengal. *Ind. Geohydrology*, VIII (2), 86-96.

Saxena, S.K., 1972: The concept of ecosystem as exemplified by the vegetation of western Rajasthan. *Vegetation*, 24(1-3): 215-227.

Tator, R.A., 1958: The aerial photograph and applied geomorphology. *Ph. Eng.*, 24, pp. 549-561.

Verstappen, H. Th., 1967: Role of landform classification in integrated surveys. *Etudes De Synthese*, VI, 35-39.

NINETEEN

G. P. Gupta

DRAINAGE SYSTEM AND FLOOD HAZARD IN BHARATPUR

INTRODUCTION

HEAVY rainfall is not the only factor which is responsible for flood. The drainage system of an area is an equally important factor. Topographically the city and its environs forms a saucer like plain, Bharatpur city being its centre. Ruprail, Banganga and Gambhir rivers create floods and bring about havoc to Bharatpur city to which there is a general slope of the country from all the three directions north, west and south. The only sluggish exit of water is in the east towards the Yamuna. This area has been in the grip of floods either every year or at least every alternate year. The research paper highlights the systematic schemes and regulations of rivers water which can save Bharatpur and its surrounding area from floods.

Flood does not mean heavy rainfall. Sometimes it is observed that one place gets higher rainfall but there is no flood and the other place which receives lower rainfall may create flood. In this way heavy rainfall is not the only factor which is responsible for flood. Drainage system of an area is an equally important factor as is the case of Bharatpur city and its environs. That minimum level of rainfall which touches and creates bad effects on the economic and other aspects of cultural landscape may be termed as flood.

Regional Setting of the City

Bharatpur city (27°13′ N; 77°30′ E) occupies a very significant location on both metre gauge and broad gauge of the Western Railway and situated at a height of 204 metres from sea level. It is also referred to as the eastern gateway of Rajasthan.

Bharatpur city is more intimately related with its environs than any other city of Rajasthan not only culturally but also physically. It is because topographically the city and its environs form a saucer-like plain, Bharatpur city being its centre. May be that is the lowest area of the State of Rajasthan. The result is that the three rivers, Ruprail, Banganga and Gambhir all non-perennial coming from its three bordering districts, Alwar, Jaipur and Sawai Madhopur respectively, have their flow from west to east steering towards this area. The first-two form an inland drainage while the third flowing east meets the Yamuna. The Banganga and the Gambhir area relatively larger rivers than the Ruprail. (Fig. 19.1).

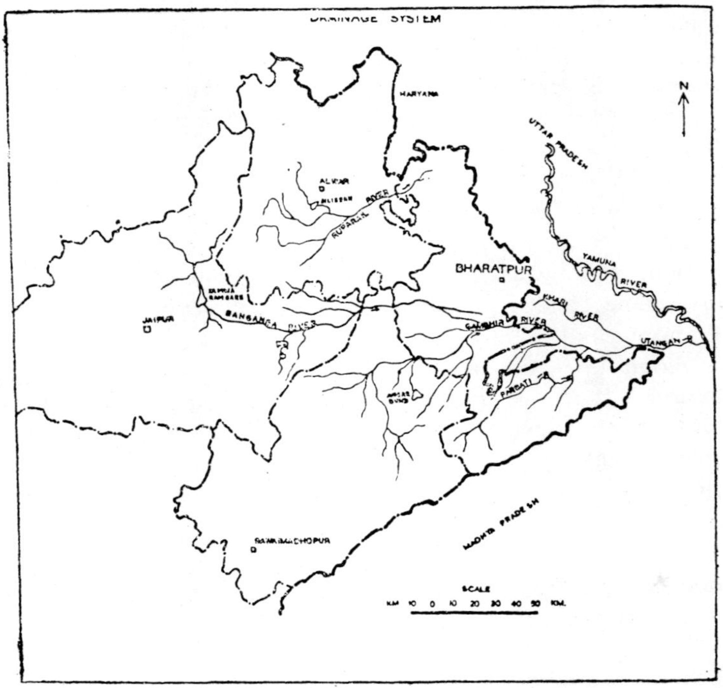

FIG. 19.1 : *Bharatpur City Environs—Drainage System*

A recent study of the Landsat imageries and aerial photographs of this area (G.S.I. Western circle) has revealed that the old channel of the Yamuna passes through the Bharatpur district.

Rainfall Trend

Since Bharatpur and its environs fall within semi-arid zone nothing untowards happens during the normal years of the monsoon. (Table 19.1) But if there is higher rainfall in the catchment of any of these rivers and that too continuously for more than 2-3 days, it creates floods and brings about havoc to Bharatpur city to which there is a general slope of the country from all the three directions—north, west and south. The only sluggish exit of water is in the east towards the Yamuna.

Since the rains are never normal but vary from year to year and from area to area, the behaviour of these rivers also becomes uncertain. With sudden heavy rains the discharge of these rivers increase and the rivers start flowing in full width restricted only with the help of marginal bunds on either banks.

TABLE 19.1 : **Rainfall in the rainy days—Bharatpur**

(In mm)

Year	June	July	August	Sept.	Annual rainfall of rainy days
1965	5.1	128.6	103.3	227.0	565.0
1972	9.1	149.0	324.7	83.9	576.6
1973	53.5	153.5	323.2	115.8	681.0
1974	46.3	363.9	151.1	36.3	621.0
1975	56.0	253.1	292.6	56.1	700.5
1976	141.1	223.8	309.5	45.5	775.0
1977	47.4	310.9	307.6	—	948.0
1978	66.2	288.4	155.2	108.0	589.9

Source: Hydrology Department of Rajasthan, Jaipur.

The hilly catchment area and seasonal nature of the rivers and the sandy alluvial nature of the soil in the environs all these factors combined together had maintained an ecological balance in the region. The hilly terrain in their upper catchment brings heavy runoff in these rivers but it had been absorbed in the sandy soils. But with the lapse of time heavy discharge during the monsoons along with heavy siltation has resulted in considerable raising of river beds.

Drainage System

Ruprail system: The Ruprail system has the following drainage system.

(a) Ujjina drain, Kama Pahari drain and Goverdhan drain complex.
(b) Homes canal.

The Ruprail river originates from the Thanagami hills, flows through the Alwar district and terminates at Sikri bund. It is sometimes responsible for creating floods in Bharatpur. At Barahweir, 20 kilometres upstream from Alwar this river bifurcates into two parts. The left side goes to Alwar and the right side leads to Sikri bund system in the Bharatpur district. The Sikri bund is 22.40 km. long and it consists of 26 regulators for canals to inundate and irrigate downstream fields. The normal flood water can be absorbed in the inundation of the system but when the flood in the Ruprail is abnormal and cannot be absorbed in the above system it starts flowing towards the Seh bund and the Motijhil towards west and north of Bharatpur city.

Whenever Ruprail is in heavy floods continuously for 2-3 days, maximum water reaches the Sikri bund and further the Motijhil through the Homes canal overflanking its banks.

The Ruprail system has been designed to drain the low lying areas in Haryana, Rajasthan and U.P. The stormy runoff of the Landoha nallah of Bharatpur, Nuh-Ujjina area of Haryana and Kama Pahari area of Rajasthan used to overflow usually during high rainfall years into the adjoining areas and ultimately drainout as sheet flow through low lying belts of Mathura and Agra in U.P. Thus the flood water used to be disposed of to the Yamuna river through western depression drain and the Khari river in Agra district. The flood water of the Sikri bund is partly disposed of through the Kama Pahari drain system and the rest via Motijhil through the Homes canal. (Fig. 19.2)

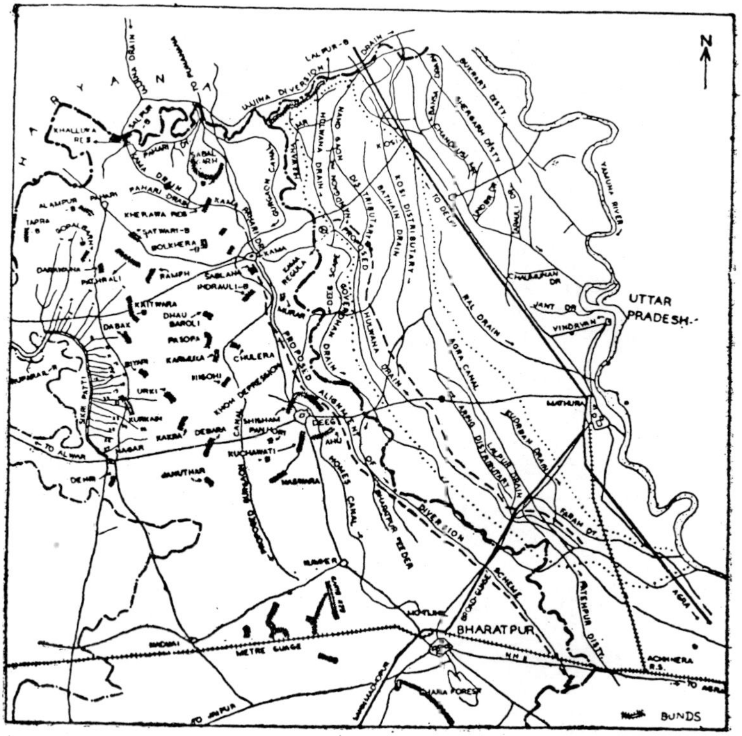

CATCHMENT AREA ......
PROPOSED AREA ---

FIG. 19.2 : *Bharatpur City North-West Environs—Integrated flood drainage Schemes of Kama Pahari Area.*

Homes canal originates from Kho depression and before discharging into Motijhil passes through the three depressions namely Kho, Ahu and Pangora with a total water spread of 7714 hactares. It passes through Gharbari, Sisham, Ahu, Soh, Borai, Bhandore and Motijhil bunds. From Motijhil the water is drained out to Chiksana bund through the Girraj canal and the escape drain. At present an unutilized water of Ruprail river goes waste and creates a flood problem in Alwar and Bharatpur districts.

Banganga System

Banganga is also a non-perennial river which originates in Jaipur district about 64 km. upstream of Jamua Ramgarh bund, after flowing 241.5 km. in Jaipur, Alwar, Sawai Madhopur and Bharatpur districts it terminates at Nekpur headworks from where the entire water of the river is diverted into Uchain canal. It enters Bharatpur district near village Kamalpura and flows for about 56 km. in Bharatpur district before entering U.P. The important tributaries of this river are Gomti and Suri. In Jaipur, Alwar and Bharatpur a number of tanks have been constructed on this river and its total catchment area up to Nekpur head is 4892.5 sq. km. (Fig. 19.3)

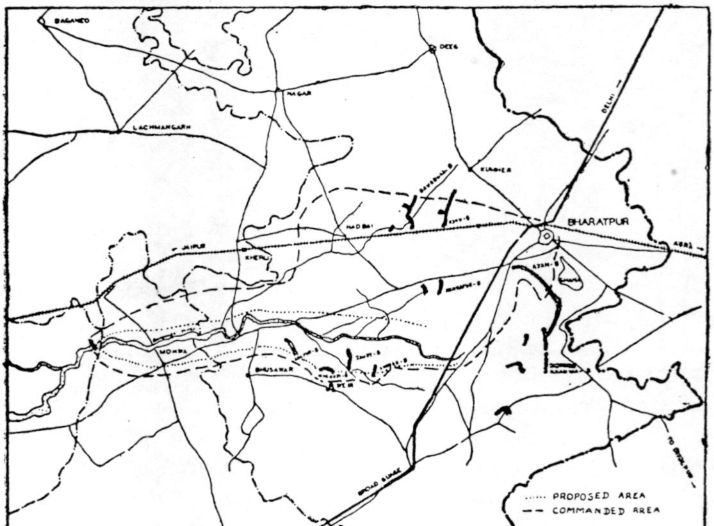

FIG. 19.3 : *Catchment Area of Banganga Area*

Gambhir System

Gambhir originates from Karoli hills in district Sawai Madhopur where it flows for 80 km. After flowing through the plains of Bayana and Rupbas tehsils of Bharatpur district it enters into U.P. where it is known as Utangan river and ultimately falls into Yamuna river. Bharatpur state constructed important tanks, several off-taking channels and the present Government has also started work in Panchna project. The total number of tanks directly fed by Gambhir system is 27 and 49 tanks excluding the Panchayat Samities tanks are fed by nallahs within its catchment area.

The Gambhir river brings large quantities of silt which is deposited along the entire course of the river when the flood recedes. In the 1972 flood which was more severe than that of the 1924, about 500 villages were submerged and 5,000 houses were washed away. A vast areas was under water and 31 irrigation tanks and canals were breached. Most of important irrigation works like the Ajan canal, Girraj canal, C.F.C.D., Chiksana canal, etc., were damaged. The Bharatpur city was surrounded with water from almost all the sides. In 1975 flood even the roads-rail traffic was badly dislocated.

Drainage around Bharatpur City

The city of Bharatpur is located in a low pocket, and has a saucer-like shape. The fortification around it consists of a huge mud-wall and a formidable moat affording protection to the garrison and the people inside the city. The solid masonry fort is also surrounded by another deep ditch called inner moat. This type of fortification was required for a town located in the depression and mudwalls constructed proved impregnable against repeated attacks. The town is located in flat plains having a very gentle slope from west to east. The elevation of the town varies from about RL 565.00 to 580.00. The water of Ruprail, Banganga and part of Gambhir systems ultimately finds access into Motijhil and Ajan bund after flowing through a series of bunds constructed upstreams. The water impounded in Ajan Bund is released through many sluices on the eastern side towards U.P. whereas water from Motijhill used to be released through Girraj canal into Chiksana bund via Noh bund. The outer moat used to get filled in from the Ajan bund, storm water of the city and surrounding areas and Motijhil through Jhula minor. The lowest invert level of the city drainage is RL 561.50 and water flows to the adjacent area when the water level in the outer moat goes beyond this level. It enters into the city near Chandpole Gate and attains a maximum RL of 570.00. This creates flooding of an area of about 4.05 hectares inside the outer mud-wall near the General Post Office. With the lapse of time the outer moat has got silted up and most of the area has been reclaimed for cultivation thus reducing the capacity of the moat which used to absorb the floods to a great extent. The surplus water of the moat flows out across the circular road or near Jaghina Gate and used to flow in sheet towards the outfall of Girraj Canal and causing heading up of levels. The surrounding of the city were thus submerged under water till the water level came down in Chiksana tank. A suitable solution to this problem was to channelise the flow from Jaghina Gate of Chiksana tank joining Girraj canal on way and demolishing Noh bund and lowering the Chiksana tank from RL 562.50 to RL 557.00 with removable wooden shutters on the top of overflow section to raise the water level in Chiksana bund up to RL 560.00 to allow inundation irrigation from the discharges received from the tanks above after passing the peak floods.

Flood Hazard Trend

Like the Bharatpur city, its environs with flat plains and many depressions has also been witnessing the havocs of the floods and the history of Bharatpur and Dholpur mentions about their rulers who had built hundreds of bunds and dams with the help of the English people in order to control floods and to irrigate extensive cultivated areas. Sikri bund, Baretha bund, Ajan bund, Sewer bund, etc. were built during the reign of Maharaja Jaswant Singh. During the same period big dams were built on the small nallahs coming down from the hilly tracts of the states. By 1904, it is learnt that within the Bharatpur state alone there were 244 small and big dams. This not only speaks about the floods havocs in the region at that time but also the ingenuity of the engineers to utilise their waters for irrigation purposes for a long time.

Although flood havoc have been for a long time in Bharatpur city and its environs, the city has been in the grip of floods either every year or at least every alternate year. During the last decade, the intensity of rainfall has shown an increasing trend with corresponding increase in the frequency of floods.

Behaviour of the Flood and the Problems

Floods caused a lot of damage to irrigation works, human life and property. Here we can recall only those years when Bharatpur town was badly affected.

There was a review of the whole situation after the severe floods of 1967 when Bharatpur town was seriously threatened by floods due to heavy discharges in Banganga and Ruprail rivers. Since the exit of water through Motijhil was limited only through Girraj canal, the water had to be disposed of during monsoons of 1967 after cutting Bharatpur-Mathura road near industrial area and cutting Bharatpur road near village Bachamdi adjacent to the district road bridge. The water which flowed across Bharatpur-Mathura road submerged the industrial area including wagon factory and overtopped the metre gauge railway line between Bharatpur Chiksana.

The floods in the Gambhir and Banganga rivers during 1972 are stated to be worst in the living memory of the people. Several villages around the Bharatpur city were submerged and 5,000 houses got washed away. The number of damaged irrigation works was about 25, the most important being Ajan canal, Girraj canal, C.F.C.D., Chiksana canal, etc. Bharatpur city was surrounded with water from almost all the sides.

During 1975 there were heavy floods in Gambhir river. This water flooded a large number of villages in surroundings of Bharatpur city.

Uncontrolled floods in Banganga river during 1977 found their way into Lower Sewar and Murwara bunds after damaging Bharatpur-Jaipur National Highway at many places. Murwara bund cut by the villagers and the flood water subsequently overtopped the broad gauge railway line between Sewar and Bharatpur stations, breached the Ajan canal behind the M.E.S. Rest House and flooded Ranjeet Nagar Colony. The Bharatpur city thus faced a pathetic situation. The water also found access into Ranjeet Nagar Colony, F.C.I. Godown and part of industrial area. The Motijhil water also overtopped Bharatpur-Mathura road and the water outflanked the regulator of the Escape drain and found direct entry into this. The water from Ranjeet Nagar Colony had to be escaped into the C.F.C.D. by cutting the Station Road near Nehru Bhawan. In short the flood situation in Bharatpur in the year 1977 presented a grim picture of chaos and misery all round.

The floods during 1978 followed almost the same route of 1977. The National Highway got washed away in a length of about 350 feet adjoining the Maloni bridge.

Measure to Control Floods

Soon after the floods of 1967, and in a meeting of State Technical Advisory Committee on 12th January 1968 following decisions were taken:

	Works proposed	Estimated cost
1.	Constructing the Escape drain from Motijhil with a capacity of 57 cumecs	Rs. 27,74,500
2.	Increasing capacity of flood control drain from Ch. 310 to 570	Rs. 13,46,200
3.	Constructing bridge across Bharatpur-Agra National Highway	Rs. 3,92,300
4.	Construction of a protection bund for Bharatpur-Mathura road	Rs. 92,300
5.	Increasing the surpassing capacity of Chiksana bund	Rs. 4,39,900
	Total	Rs. 50,45,200

The estimates of Escape drain with a carrying capacity of 57 cumecs amounting to Rs. 33 lacs including the provisions of two railway crossings on Delhi-Bombay broad gauge and Jaipur-Agra metre gauge line was finalised in July 1968.

The system as partly completed had the following drawbacks :

1. An integrated scheme for the escape drain to carry 57 cumecs from Motijhil to the outfall at Kheri river was not prepared with respect to heading up on account of high F.R.L. in the submergence of Chiksana bund.

2. The storage capacity of the outer moat was not restored after doing proper excavation.

3. Due to high level in the C.F.C.D. the city inlets did not function adequately necessitating lowering of the outfall level and deepening of the outer moat of the drain.

4. The escape drain was ineffective for want of construction of the two railway bridges.

5. The widening of the drain from Ch. 120 to Ch. 206 was not included in the original estimate.

6. The discharging capacity of Girraj canal is only 17 cumecs and its further widening or remodelling of the structures is not possible as it passes through populated area. As such the capacity of the Escape drain had to be correspondingly increased from 57 cumecs to 68 cumecs so that the total discharge from the Motijhil is of the order of 85 cumecs. Due to paucity of funds much work could not be done as approved by the State Flood Control Board.

The Scheme was once again reviewed in 1971.

The proposals finalised during 1976 comprised of draining off all the surplus water of Bharatpur and its surroundings and to deplete all the low pockets and its safe disposal without causing any damage on way. After observations of the monsoons and assessment of floods witnessed during 1977 the discharge figures were again revised and subsequently approved by the Chief Engineer Irrigation, Rajasthan, Jaipur. The figures of discharges as incorporated in different projects reports are shown below:

(In cumecs)

	1964	1968	1976	1977
1. Escape drain	—	57	68	102
2. Girraj Canal	28	28	17	23
3. C.F.C.D.	11	11	11	11
4. Ranjeet Nagar drain	—	—	—	34
Total	39	96	96	170

The latest proposals were incorporated after seeing the floods of 1977 as mentioned above. The cost of the new proposal works out to about Rs. 235 lacs as against Rs. 298 provided in the revised project estimates of 1976.

The yearwise expenditure on the C.F.C.D. main drain complex is given below:

S. No.	Year	Expenditure
1	1964—65	2,68,020
2	1965—66	7,64,486
3	1966—67	4,52,263
4	1967—68	3,20,296
5	1968—69	15,74,264
6	1969—70	12,48,281
7	1970—71	7,31,228
8	1971—72	49,207
9	1972—73	3,83,012
10	1973—74	7,09,266
11	1974—75	18,57,696
12	1975—76	5,21,020
13	1976—77	25,78,135
14	1977—78	23,40,066
15	1978—79	40,41,600
	Total	1,81,26,041

Ranjeet Nagar drain: The present Ranjeet Nagar Colony is situated in a low depression and there is no exit for the water when the area is inundated by breaches in Motijhil or Ajan canal as happened in 1967 and 1977 or by local incessant rains. During the visit of Central Study Team after the monsoon in 1977, it was stressed that this accumulated water should be confined in a drain and disposed of suitably. This drain was designed for 8.50 cumecs or equal to the discharge of the Ajan canal and called the Ranjeet Nagar drain which is under construction. It will cost about Rs. 25 lacs.

Thus it may be seen that whenever there is any excess water in any of the three rivers it rushes towards Motijhil and Bharatpur town. Also it is observed that floods had generally come during the month of August only. To save the Bharatpur town from floods it is very essential to keep Motijhil empty till the end of August. In that case if there are heavy floods in any of the above rivers, Motijhil will help a lot in absorbing this flood and passing it through Girraj canal and flood drain slowly in Chiksana bund and finally in the Khari river or Utangang river in U.P.

The capacity of the Motijhil is 235 m. cft. or 2730 cumecs days and the capacity of Girraj canal is 600 cusecs. The flood water of these rivers reaches up to Motijhil slowly after fllling all the bunds in its way. Floods of one river can easily be absorbed and passed away through Girraj canal. In case when the duration of heavy floods remains more than 2-3 days in a river or it comes simultaneously in two rivers, the system cannot absorb the flood water and danger of flooding Bharatpur town is created. To overcome such situation anti-flood drainage scheme of Bharatpur is under construction. In this scheme deepening of outer moat construction of a flood drain of 2000 cusecs are included.

It is hoped that after completion of this anti-flood drain scheme there will not be any danger to Bharatpur city from flood. But at the same time proper and timely regulation of all the river water is very essential. Only a systematic regulation of river water as discussed above can save Bharatpur and its surrounding area from floods as well as from drought where it is necessary to keep the floods away it is very essential to provide irrigation by flooding the field timely and to utilise this unstored river water.

This flood water is a boon also for Bharatpur as it produces bumper crop and silt carried by Banganga river is said to be very good manure.

SUGGESTED READINGS

Buddi Ram, 1974: Flood problems in Bharatpur District, M.A. Dissertation, University of Rajasthan, Jaipur, 1974.

Gupta, G.P. 1980: Cultural landscape of Bharatpur City and its environs, (an unpublished Ph. D. Thesis, University of Rajasthan Jaipur, 1980.

Jaswant Singh, 1896: History of Bharatpur Abu collection, University of Rajasthan, Jaipur.

Kumar, P.N., 1968: Flood protection method in India, *Geographical Review of India*, Calcutta, Vol. XXX, 1968.

Sehegal, K.K., 1971: *Rajasthan District Gazetteers*, Bharatpur, Govt. of Rajasthan.

Reports

Master Plan for Flood Control Works in Bharatpur District, Vol. I and Vol. II, 1979, Flood Department, Bharatpur.

INDEX